Designing the Coal Preparation Plant of the Future

Edited by:
Barbara J. Arnold
Mark S. Klima
Peter J. Bethell

Published by

Society for Mining,
Metallurgy, and Exploration, Inc.

Society for Mining, Metallurgy, and Exploration, Inc. (SME)
8307 Shaffer Parkway
Littleton, Colorado, USA 80127
(303) 973-9550 / (800) 763-3132
www.smenet.org

SME advances the worldwide mining and minerals community through information exchange and professional development. SME is the world's largest association of mining and minerals professionals.

ISBN-13: 978-0-87335-257-4

Library of Congress Cataloging-in-Publication Data

Designing the coal preparation plant of the future / edited by Barbara J. Arnold, Mark S. Klima, Peter J. Bethell. -- 1st ed.
p. cm.
Includes bibliographical references and index.
ISBN 978-0-87335-257-4
1. Coal preparation plants--Designing and construction. I. Arnold, Barbara J. II. Klima, Mark S. III. Bethell, Peter J.

TH4561.D47 2007
662.6'23--dc22

2007040123

Contents

Contributors

B.J. Arnold
PrepTech, Inc.
Apollo, Pennsylvania

Christopher J. Barbee
Arch Coal, Inc.
Charleston, West Virginia

Peter J. Bethell
Arch Coal, Inc.
Charleston, West Virginia

Paul Brodzik
Derrick Corporation
Buffalo, New York

David M. Carris
John T. Boyd Company
Canonsburg, Pennsylvania

Van L. Davis Jr.
Alpha Natural Resources
Beckley, West Virginia

Greg DeHart
Cardinal Plant
Mountain Laurel Complex
Arch Coal, Inc.
Sharples, West Virginia

G.J. de Korte
CSIR Miningtek
Pretoria, South Africa

S.W. Eck
Pennsylvania Crusher Corporation
Broomall, Pennsylvania

J. Engelbrecht
Multotec Process Equipment
Johannesburg, South Africa

B. Firth
CSIRO Energy Technology
Queensland, Australia

Rick Q. Honaker
University of Kentucky
Department of Mining Engineering
Lexington, Kentucky

Mark S. Klima
Pennsylvania State University
Department of Energy and Mineral Engineering
University Park, Pennsylvania

J.N. Kohmuench
Eriez Manufacturing
Erie, Pennsylvania

Gerald H. Luttrell
Virginia Tech
Department of Mining & Minerals Engineering
Blacksburg, Virginia

M.J. Mankosa
Eriez Manufacturing
Erie, Pennsylvania

M. O'Brien
CSIRO Energy Technology
Queensland, Australia

Thomas J. Porterfield
Farnham & Pfile Engineering, Inc.
Monessen, Pennsylvania

Fred L. Stanley
Alpha Natural Resources
Abingdon, Virginia

Todd Vander Hill
BV3D
Monessen, Pennsylvania

Richard Woodward
Thermo Fisher Scientific
San Diego, California

Giancarlo F. Yannoulis
Humboldt Wedag Australia Pty. Ltd.
New South Wales, Australia

Dieter Ziaja
Humboldt Wedag
Coal & Minerals Technology GmbH
Cologne, Germany

Preface

Although much has been written about the "nuts and bolts" of coal processing, considerably less has been published about the industry from a designer's or engineer's point of view.

Designing the Coal Preparation Plant of the Future is intended to fill that void.

We asked more than a dozen of the world's foremost experts to share their insights with us.

These 15 informative, meticulously researched chapters provide a compelling road map of where we've been and where we need to go, what we're doing today, and, most importantly, how we can do it better.

The editors thank these authors for their valuable contributions. We believe *Designing the Coal Preparation Plant of the Future* is truly groundbreaking work for an industry where groundbreaking is a long-standing, proud tradition.

Barbara J. Arnold, PrepTech, Inc.
Peter J. Bethell, Arch Coal, Inc.
Mark S. Klima, The Pennsylvania State University

PART 1

Where We've Been and Where We Need to Go

A Historic Perspective

David M. Carris

ABSTRACT

Coal preparation has been practiced in the United States for well over a hundred years. During this period, the coal industry has experienced many equipment and technology advances. This chapter discusses some of the key process developments that have occurred.

INTRODUCTION

Coal preparation had its roots in the United States in the late 1800s. The anthracite field in eastern Pennsylvania was a major coal producer, providing fuel for home heating, stokers, and metallurgical uses. Coal was mined by hand and the slate material separated at the underground working face before it was loaded into a mine car. This gives meaning to the saying "coal preparation begins at the face." Coal operators in the bituminous fields were providing coal for various home, industrial, and other thermal coal markets and for use in blast furnaces. Coal that could be mined cleanly was crushed and screened into discrete sizes such as egg, stove, net, and pea for the various markets. If the coal seam could not be mined cleanly or contained parting material, it would need mechanical washing. Typically, coal was crushed and screened to the various market sizes before it was washed in a cleaning device. The anthracite industry commonly called the plants "breakers," and the bituminous industry called them "tipples," which are terms carried over from Europe.

1850–1899

Most early process equipment was developed in Europe in the mid- to late 1800s. Some examples of this early equipment include

- Roll crushers (1840)
- Rheolaveur launder (1840)

- Bessemer chloride washer (1859)
- Revolving screens (1860)
- Bradford breaker and picking table (1873)
- Shaking screens

In 1892, the first steam-powered wet breaker plant was commissioned in the anthracite industry. Up to this time, crushing, screening, and hand picking were the principal methods of preparing coal. Wet cleaning was gaining acceptance in Europe with the Robinson classifier and centrifuge process plant. In 1892, the first Robinson–Ramsey plant was installed at Tennessee Coal, Iron and Railroad Company, and by 1912, eight new classifier plants had been placed in operation. By 1890, shaking screens were being used in plants to size coal. Wet processing plants were gaining acceptance in the coal industry as a viable means to clean coal.

1900–1949

By the turn of the century, many manufacturing companies were pursuing the development of coal process equipment for both dry and wet separations. Underground mining practices were changing, with drilling and blasting of coal followed by conventional mining. There was an increase in impurities in the run-of-mine (ROM) coal with this mining process, thus increasing the need to clean the coal on the surface in a processing plant.

In 1916, the American Coal Cleaning Company in Welch, West Virginia, built its first air separating plant. Air cleaning gained popularity with the development of the Peal Davis machine (1924) and the Stump Air-Flow jig (1932). By 1940, air machines processed 12 million tons of clean coal. In 1941, Roberts & Schaefer Company introduced the Super Air-Flow table, which defined air cleaning of coal. Typically, air separation was used to process the ½-in. × 0.6-mm (28-mesh) size fraction. In the 1940s, dedusting of coal to remove the minus-1-mm (16-mesh) size fraction ahead of wet processing became popular.

Wet processing of coal was rapidly being developed as a more efficient method to remove the impurities from the ROM coal. Coal buyers were increasing their demands for better product quality. A number of coarse coal wet separating processes emerged during this period as follows:

- Chance cone dense-media separator (1917)
- Trent hydrotator separator (1920s)
- Menzie cone separator (1924)
- Baum jig washer (1930s)
- Belknap chloride dense-media washer (1935)

- Tromp dense-media washer (1938)
- Barvoy dense-media washer (1940)

The first hydrotator plant was constructed by Wilmont Engineering in 1939 to clean 2-in. × ⅜-in. anthracite coal by individual size fractions. By 1941, there were 121 Menzie cone separators in operation in the anthracite industry, with a total washing capacity of 9,784 tph. In 1940, American Cyanamid installed the first dense-medium cone separator. By 1945, Dutch State Mines had placed a 15-tph dense-medium pilot plant in operation, using magnetite as the media.

Fine coal processing developments included Deister Overstrom concentrating tables (1920) and Rheolaveur launders (1924) for mid-size coal. In the late 1930s to early 1940s, fine coal equipment such as vibratory screens, drums, disc vacuum filters, and centrifugal dryers were being introduced and used in the preparation plant processes. Thermal drying of coal using rotary dryers, multilouver dryers, and Raymond flash dryers was used to reduce the moisture in coal fines. Froth flotation (1918) was used sparingly because of its high cost.

The early plants were made of wood and were constructed over the rail loading tracks so that coal could be loaded directly into railcars by commercial and industrial market sizes. As the plant capacities increased and processes became more complex, steel structures were used to handle the increased loading. By 1949, there were 571 preparation plants operating, with clean coal production reaching 153.6 million tons per annum.

1950–1979

Coal preparation in the United States underwent a substantial evolution when the Dutch State Mines technology was introduced in the late 1950s. The first dense-medium (magnetite) cyclone plant was commissioned in 1961, and Heyl & Patterson had two 50-tph pilot plants in operation by 1965. By this time, static sieve bends and classifying cyclones were being used in the plant process circuits.

Air separation of coal reached its peak in 1965 when approximately 25 million clean tons were produced using this technology. The largest air plant had a 1,500-tph capacity and was operated by the Florence Mining Company in Pennsylvania.

From 1938 to 1964 coal processed in jig washers increased from 27 million tons per year to 146 million tons per year. These units were supplied by companies such as Jeffrey, Link-Belt, and McNally-Pittsburg. Jig washers were the principal method used to clean coarse coal during this period.

In the 1960s, dense-medium processes were used in plants that produced a metallurgical product because of their ability to clean at a separating gravity ranging from 1.30 specific gravity (SG) to 1.80 SG. Deister tables typically were used to process the ¼-in. × 28-mesh size fraction and flotation used for the finer size coal. Most metallurgical plants also installed thermal drying equipment to reduce the plant product moistures. Fluid bed drying systems became the preferred method to dry coal thermally. Approximately 481 plants were operating in 1969 because of the replacement of smaller and obsolete plants in this 20-year period.

The 1970s saw a dramatic increase in coal demand and an accompanying surge in coal preparation plant construction. Mines were becoming larger, with increases in capacity made possible by the expanded use of continuous miners and longwall systems. Large utilities' demands for better heating value of fuel and lower sulfur in the coal spurred the development of new mines and plants. Plant capacity rates dramatically increased, with many in the 1,000- to 2,500-tph range. Plants constructed in the 1940s and 1950s were being replaced because they were obsolete and could not make the desired product quality. Plant engineering designs were innovative and ranged from all gravity feed circuits (with plant structures reaching +55 m [+180 ft] in height) to low-profile mineral circuit design (18 m [60 ft] in height). Modular plants in the capacity range of 100–300 tph became popular, using dense-medium cyclone cleaning-to-zero technology.

Plants were no longer being constructed for loading directly into railcars. Clean coal handling, storage, and railcar flood loading stations were being installed at both steam coal and metallurgical coal mines. In 1977, federal regulations for noise, dust, and the disposal of fine refuse material affected plant design. By the end of the 1970s, approximately 519 plants were operating.

1980–PRESENT

The 1980s saw the introduction of automation in plant electrical design. The first Allen-Bradley programmable logic control (PLC) system was installed in a plant in Ohio. PLC allowed all the material handling and plant process equipment to be controlled by one operator, thus reducing the plant's workforce needs. On-line ash and moisture analyzers were in development to give operators real-time information as they processed the coal. This was clearly the age of development of wear materials, equipment, and process circuits. Ceramics and high-density plastics were being tried in numerous applications and in equipment to reduce wear and plugging. Spiral separators were making a return to process fine coal because of the better materials used to manufacture the units. Column flotation was in development, and oil agglomeration circuits were being tried in plants to reduce pyrite and recover fine carbon. In order to accommodate increased mine output, driven largely by increased use of longwall systems, plant throughput rates were increased.

Increased production needs forced plants to operate 18 shifts or more per week to handle the mine production. Plant operators had to change their mode of operation from the conventional two-shift operation and one maintenance shift per day to running the plant continuously during the week with maintenance on the weekend. It was also critical that plants maintained >90% mechanical availability to keep pace with mine production. By the end of 1989, approximately 557 plants were processing coal.

In the 1990s, the coal industry consolidated, and many smaller mines closed because they were unable to compete with the larger operations. Coal preparation plants were being required to blend and process multiple seams at higher capacity rates. At the same time, the quality of the coal seams being mined decreased, resulting in lower clean coal yields. Plant process rates were on the rise, with many plant facilities being constructed with a 1,000- to 3,000-tph capacity rate. Plant process equipment capacity became larger with the development of banana screens and large-diameter dense-medium cyclones, thus lowering plant capital costs per ton-hour. Fine coal process circuits improved with the use of column flotation and screen bowl centrifuges to recover the ultrafine size coal. On-line coal analyzers provided the plant operators with a method to maximize product recovery and meet product quality specifications. By the end of 1999, though the number of operating coal preparation plants had decreased to approximately 212 from a high of 557, the tons of coal processed steadily increased during this period.

The stagnant coal industry of the 1980s and 1990s showed few new mines being developed and hardly any new coal preparation plants being constructed. Reduced demand for coal preparation construction saw a reduction in the number of companies providing these services. Companies such as Heyl & Patterson, J.O. Lively, and McNally-Pittsburg disappeared. Today, only a few large coal preparation design and construction firms remain in business.

From 2000 to 2007, dense-medium cyclones, spirals, and froth flotation process circuits dominated new plant construction because of their lower capital cost per ton-hour. Beneficiation of lower-rank coal through air separators or thermal dryers is being used to separate impurities and improve the heating value of the coal. Plants continue to increase in capacity, with the Bailey plant facility leading with more than 6,000 tph feed rate capacity. The United States has the highest process capacity in the world, with the ability to clean more than 636 million tons per year (in 2001).

FUTURE

The resurgence in coal prices across the United States in the 2003–2006 period resulted in increased coal demand. While existing mines attempted to expand and many new mine development projects were announced, negative impacts emerged in terms of labor shortages and increased costs for steel, equipment, and third-party services. These cost pressures also affected coal preparation.

In the past 2 years, the capital costs for constructing a coal preparation plant have more than doubled. This makes it difficult to justify greenfield mines and preparation plant facilities. In addition, the industry faces the challenge of developing improved methods for disposing of fine refuse material that meet state and federal regulations.

As coal mining continues its progression, mining operations, particularly in Appalachia (eastern United States), will face less favorable coal seams and more difficult mining conditions. The next generation of coal preparation engineers will face these new challenges.

SUPPLEMENTAL REFERENCES

Keystone Coal Buyers Manual (annual publication). New York: McGraw-Hill.

Mitchell, D.R., ed. 1943. *Coal Preparation*. Steely W. Mudd Series. New York: American Institute of Mining and Metallurgical Engineers.

Proceedings of the International Coal Preparation Exhibition and Conference (various years, 1984–2007), Lexington, KY. Stamford, CT: Primedia Business Exhibitions.

Prochaska, E. *Coal Washing*. New York: McGraw-Hill.

Today's Coal Preparation Plant: A Global Perspective

Peter J. Bethell and Christopher J. Barbee

ABSTRACT

This chapter describes coal processing circuitry and practices used worldwide. Coal sizing, cleaning, dewatering, and reject disposal devices, techniques, and practices are discussed, with regional variations highlighted. Typical flowsheets for plants in major coal processing regions are included, with differences between countries explained.

SIZING

Raw Coal

To achieve optimum separation of coal from waste material in a plant, several cleaning devices are used. Each of these treats only a narrow size fraction efficiently (Figure 1). Consequently, run-of-mine coal entering a plant must be separated to present appropriately sized material to each cleaning process. Typically in the United States, coal is divided into four separate size factions for cleaning: coarse, 6 in. × ½ in. nominal; intermediate, ½ in. × 1 mm nominal; fines, 1 mm × 0.150 mm (100 mesh [100M]); and ultrafines, –100M.

Separation at ½ in. is becoming more common with the use of multislope high-capacity banana screens. Typically, U.S. screens are up to 10 ft wide × 20 ft long. These units separate efficiently and have far greater capacity than the traditional combination single-slope raw coal and horizontal pre-wet screen combination. Punch plate is the conventional decking material used for this application, but an increasing percentage of polyurethane panels are being used for their superior wear characteristics.

Usually new plants also use banana screens for separation at 1 mm, although many plants still use a combination of sieve bends and horizontal screens. Australian and Chinese plants, which usually use single-deck screens,

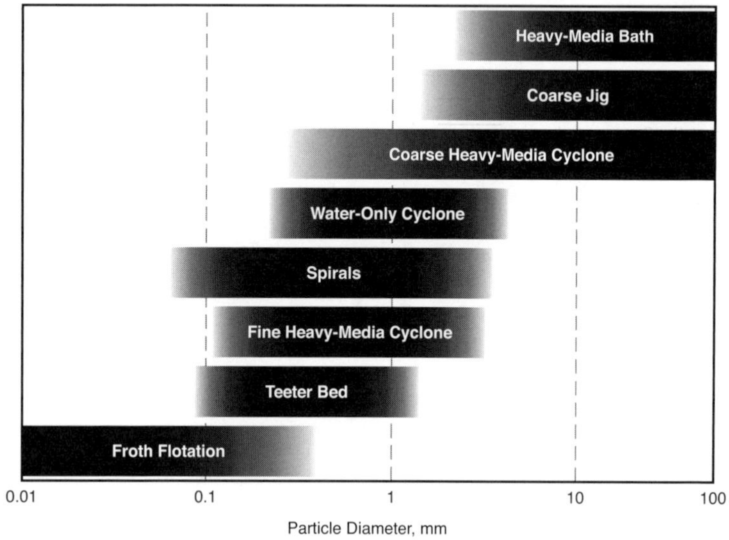

FIGURE 1 Effective sizing for coal cleaning efficiency, by process

use 12-ft- and 14-ft-wide banana screens. However, plants in the United States, where the more structurally complex double-deck screens are more common, stop at 10-ft-wide screens. Deck material used for making the 1-mm separations can be profile steel wire or polyurethane. In some applications, the profile wire is chrome plated to extend the life of the screen panel.

Several U.S. plants now use a combined raw coal and deslime double-deck banana screen. The top deck feeds the dense-medium vessel + ½ in., and the bottom deck feeds the dense-medium cyclone circuit ½ in. × 1 mm, with –1 mm passing to the fines circuits (Bethell and DeHart 2006).

Separation of fines from ultrafines (1 mm × 100M from –100M) invariably is achieved using classifying cyclones. Cyclone diameters typically run in the 15- to 20-in. range. However, where coal frothability is good, larger-diameter (up to 36 in.) cyclones have been used to make a coarser cut (~0.25 mm). The smaller-diameter units typically are used where froth flotation is not practiced and cyclone overflow is discarded (smaller-diameter cyclones make a finer cut). A schematic of a typical U.S. central Appalachian plant with sizing, separation, and dewatering devices is shown in Figure 2 (Bethell et al. 2006).

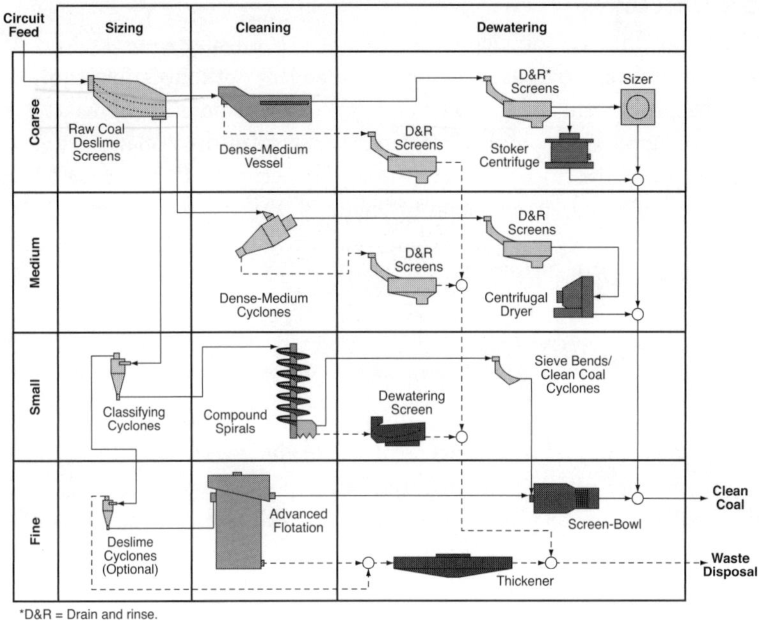

FIGURE 2 Typical U.S. central Appalachian coal processing plant

Fine Coal Desliming

After the 1 mm × 100M coal has been cleaned, it still carries undesirable slimes, which must be removed. This can best be achieved by two-stage fine wire sieves (Bethell and DeHart 2004) with intermediate repulping or by classifying cyclones followed by fine wire sieving. Derrick Manufacturing has recently introduced a stack sizer that also performs this function (Hollis 2007).

Media Recovery Circuitry

Typically horizontal screens are used for drain-and-rinse (D&R) applications in vessel circuits. Dense-medium cyclone media recovery circuitry uses both banana screens and sieves with horizontal screens. Recently, the incorporation of a flat flume screen ahead of D&R banana screens has shown excellent results in enhancing magnetite recovery (Metzler 2007). The advent of the high-gauss self-leveling magnetic separator has greatly facilitated magnetite recovery.

Some plants incorporate magnetite thickening or overdense systems. However, where plant operating densities are not expected to vary widely, feeding dry magnetite into the dense-medium sumps proves most cost-effective. Density control through water injection into the dense-medium suction is widely practiced using magnetite addition to make up sump levels.

Roller Screens

Where raw fines are sufficiently low in ash, it is possible to back-blend this material with washed coarse fractions. The tendency of conventional horizontal or banana screens to blind when dry-screening damp material makes their use in this application problematic. Consol Energy, at the Robinson Run and McElroy plants, has efficiently used Roxon roller screens to dry-screen at ±⅜ in. with minimal blinding, as shown in Figure 3 (Yanchak and Metzler 2006). The same technology has also been applied successfully at Arch Coal's Castle Valley plant in Utah.

COAL CLEANING

Coarse

For the most part, U.S. coal plants use dense-medium vessels for separating the +½-in. material, primarily those of the bath type. Some plants use jigs, and a few (smaller plants or plants processing coals with minimal coarse material) process coarse (crushed to –2½-in.) and intermediate-sized material in a single dense-medium cyclone circuit. The dense-medium bath provides good separation efficiency at low capital and operating costs, generating fewer fines than the dense-medium cyclone circuits, thereby providing lower moisture content and higher steam coal plant yield.

FIGURE 3 Roxon roller screen at McElroy plant

South Africa, with more near-gravity material in Gondwana-type coals, typically uses the slightly more efficient drum (e.g., Wemco, Teska) separators for coarse coal cleaning. However, recent circuit designs in South Africa have favored coarse (3 in. × ⅜ in.) and fine dense-medium cyclone circuits (⅜ in. × 1 mm; Cresswell and Salter 2006).

Australian plants generally use combined intermediate and coarse coal cleaning in dense-medium cyclones. Australian metallurgical coal supply contracts have less stringent moisture specifications. Also, the Australian climate is more amenable to natural evaporation (very little rain, hot dry summers, and winters with no freezing days). Coupled with some coal liberation brought about by crushing, these factors have led to the dense-medium cyclone circuit replacing the U.S. dense-medium vessel and dense-medium cyclone approaches (Bethell et al. 2006).

Intermediate

Efficient separation of nominal ½ in. × 1 mm coal is achieved almost completely by dense-medium cyclones in the United States. With improved magnetite recovery circuitry (self-leveling high-gauss separators) and improved screen capacities, capital costs have been reduced.

Cyclone diameters and capacities have continued to increase to the point that 5-ft-diameter units capable of handling 500 tph are in service in Australia, as shown in Figure 4. This helps reduce plant capital costs and simplifies circuitry.

Whereas separation in these units down to ⅛ in. is efficient, performance in the fines fractions is poorer, particularly in units treating large-topsize material. This phenomenon explains why the South Africans prefer the two-dense-medium cyclone circuit approach (large-diameter dense-medium cyclone for +⅜ in., smaller-diameter dense-medium cyclone for ⅜ in. × 1 mm).

With most U.S. plants incorporating a dense-medium vessel, the need for massive cyclones is reduced. Also, in the United States some plants are incorporating a modular single-unit operation philosophy (single raw coal and deslime screen, singe dense-medium cyclone, and single clean coal and refuse screens; Bethell and DeHart 2006). The dense-medium cyclone that matches this concept typically is less than 40 in. in diameter.

Fines

During the last 10 years, spiral circuits have become the predominant fine cleaning device in the United States. Typically a combination of water-only cyclones with spiral recleaning of water-only rejects or a standalone spiral circuit is used.

To achieve maximum efficiency in spiral circuits, recleaning of the primary spiral middling or the use of a compound (middling or clean coal retreat) is needed, as shown in Figure 5. Low loading of the spiral (2–2.5 tph/start) also

FIGURE 4 Large-diameter Minco Australian cyclone

FIGURE 5 Compound spiral showing repulping box

maximizes efficiency and reduces the cutpoint into the 1.70–1.80 specific gravity range (Luttrell et al. 2003). The units have no moving parts and, if preceded by an adequate protection screen and fed consistent flow, will perform consistently well.

With the need in some plants to make a lower-gravity separation than possible in a spiral, some Australian plants are using teetered-bed separators to treat fine material. These units, though of high capacity, treat the coarse material at a very low specific gravity (1.40), with the fines being treated at a much higher specific gravity (2.0+). This is contrary to plant yield optimization, which is best achieved when all size fractions in a plant are separated at the same density.

Canada typically uses water-washing cyclones to treat 1 mm × 100M material. South African plants typically use spirals for fines, although some teetered-bed separator testing is proceeding.

Ultrafines

Treatment of ultrafines varies from total discard to flotation of all –100M fractions. An emerging trend, the treatment of only the +325M material, is gaining strength from both steam and metallurgical coal plants (Baumgarth et al. 2005). For maximum recovery, flotation of –100M material should be accompanied by filtration. However, without thermal drying this typically delivers high moisture levels, which may often render the total product out of shippable specification limits.

The most widely used circuitry in modern U.S. plants is shown as Figure 6. Desliming at ~325M is followed by column flotation of the +325M material to achieve maximum selectivity and reject the most clay slimes. The deslimed column product will be dewatered in a screen-bowl centrifuge with spiral product, and moisture content typically will be low (~10% surface moisture). The use of a screen-bowl centrifuge will reduce moisture content, but up to 50% of the –325M product will be lost.

Australian coal plants have taken a different approach. For the steam plants of the Hunter Valley, –100M slimes typically are discarded. The 3 in. × 1 mm material is treated in large-diameter dense-medium cyclones. Teetered-bed separation or spiral circuits are used on the 1 mm × 100M material. Many Australian metallurgical coal plants traditionally have incorporated two circuits: large-diameter dense-medium cyclones for +0.5 mm and froth flotation for –0.5 mm, as shown in Figure 7. Conventional froth flotation has been superseded by column flotation (Jameson and Microcell; Stone et al. 1995). The column circuitry's superior selectivity has produced lower-ash products, thereby allowing dense-medium cyclone operating gravities to be raised, maximizing overall plant yield. The loss of the coarse material (misplaced +0.5 mm)

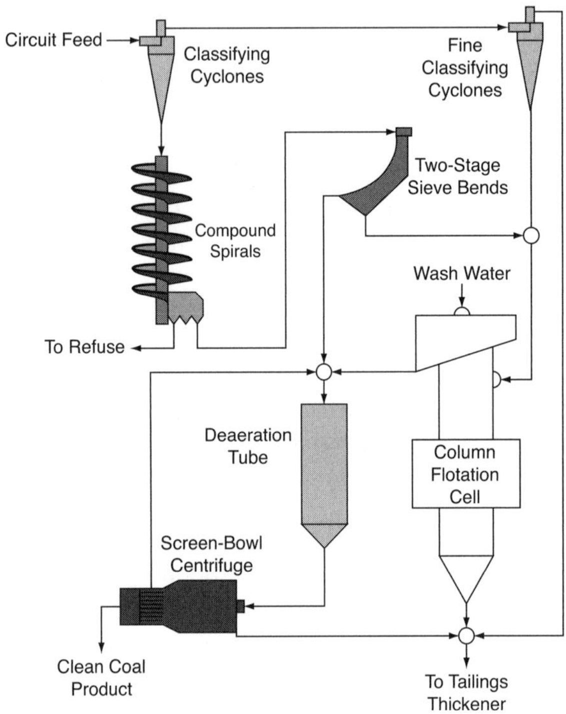

FIGURE 6 Deslime column flotation circuit

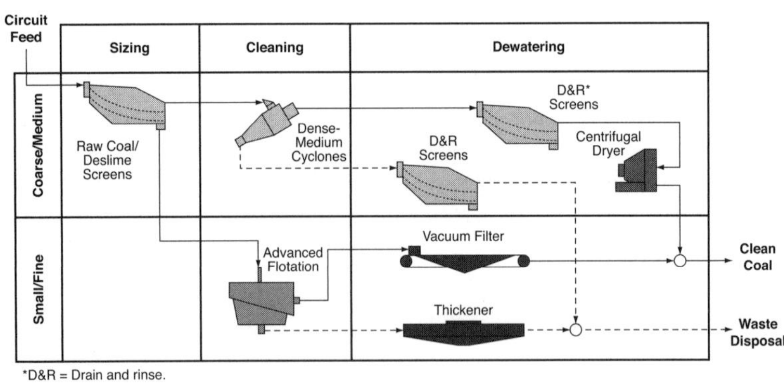

FIGURE 7 Traditional Australian Queensland two-circuit metallurgical coal plant

in the froth tails has led to the inclusion of a third circuit (spirals) in several new installations (Gray 2007).

Canadian coals tend to be very friable, generating copious fines. Typically Canadian plants handle these –100M fines in column or conventional flotation cells with filtration or solid-bowl centrifuges. Thermal drying in many instances is necessary because of the high volume of low-ash fines in the coals (Hogg 2007).

In China, dry coal cleaning is widely used in upgrading marginal direct-ship coals. The air table is widely used in this capacity (Honaker et al. 2007).

DEWATERING

Typically U.S. plants do not centrifuge +½ in. material, sending clean coal screen product directly to the clean coal conveyor. The size class ½ in. × 1 mm normally is dewatered in a vibrating or screen-scroll centrifuge. The latter typically produces lower moisture but takes more maintenance time.

New designs of horizontal vibrating centrifuges are gaining wide acceptance in the United States. These units typically have high capacity (+200 tph) and far better maintenance characteristics than either of the vertical machines.

Fine coal centrifuges (screen-scroll type) are still used for fine coal dewatering. However, most new plants use screen-bowl centrifuges for dewatering fines or fines and ultrafines. The screen-bowl centrifuge offers superior coal recovery to the fine coal centrifuge and reduced moisture content. The large screen bowls (44 in. × 132 in.) handle 80 tph (dry solids) and 800 gpm of slurry. Typically a combination of spiral product and deslimed froth concentrate produces an ideal screen-bowl centrifuge feed.

Filtration, which has superior slime recovery compared with screen-bowl centrifuges (95% or more in the filter vs. 50%–75% in the screen-bowl centrifuge), is still used in the United States, primarily when backed up by thermal drying.

Most Australian plants that froth coal use either drum, disc, or more recently horizontal belt vacuum filtration, as shown in Figure 8 (Bethell 2005). Horizontal belt vacuum filtration is also used in South African plants such as Grootegeluk to dewater fines. Where higher-moisture products can be tolerated, filtration provides superior coal recovery.

REJECT DISPOSAL

Coarse Refuse

Coarse plant refuse is typically trucked or belted to an approved coarse refuse site where it is contoured and stabilized. Some U.S. plants form a combined coarse and fine refuse product, with the fines generated from thickener

FIGURE 8 Horizontal belt vacuum filter

underflow treated by belt presses. Similar practices occur worldwide. Some total reject pump-away is practiced (high-yield plants). Total plant rejects are slurried and pumped to a tailing site where settlement takes place.

Fine Refuse

In the United States, fine refuse disposal typically occurs preferentially in impoundments. Thickened tailings are pumped behind a dam built of coarse refuse, which is covered and reclaimed at project completion. Pumping tailings into abandoned underground workings is also practiced whenever possible. If those alternatives are not practicable, slurry cells may be used to dispose of coarse and fine refuse.

As a last resort, belt filters may be used. These are high-maintenance, high operating cost units that consume copious volumes of flocculent. The product from these units is also hard to handle, typically at +45% moisture. Paste thickening is beginning to be practiced on coal tailings as a lower operating cost alternative to belt presses (Henry 2007). The paste thickener at Arch Coal's Lone Mountain plant is shown in Figure 9.

Impoundments are unavailable in parts of Europe, Australia, South Africa, and China because of space limitations or environmental constraints. In these countries, the use of hyperbaric disc filters or plate-and-frame pressure filters is now fairly common.

FIGURE 9 Paste thickener

REFERENCES

Baumgarth, T., Bethell, P., and Gupta, B. 2005. Recovering an additional 20 T.P.H. of coal through a deslime column flotation circuit addition at Lone Mountain Processing–Virginia. In *Proceedings of Coal Prep AGG 2005*, Lexington, KY. Stamford, CT: Penton Media. pp. 39–50.

Bethell, P. 2005. Current worldwide coal cleaning equipment, circuitry, and practices. In *Proceedings of Coal Prep AGG 2005*, Lexington, KY. Stamford, CT: Penton Media.

Bethell, P., and DeHart, G. 2004. Fine circuit upgrades at Hobet (Beth Station) plant. In *Proceedings of Coal Prep AGG 2004*, Lexington, KY. Stamford, CT: Penton Media. pp. 37–49.

———. 2006. The design, construction, and commissioning of the new 2000 TPH Arch Coal Cardinal preparation plant. In *Proceedings of the XV International Coal Preparation Congress and Exhibition*. Beijing: China National Coal Association. pp. 79–88.

Bethell, P., Luttrell, G., and Firth, B. 2006. Comparison of preparation plant design practices for metallurgical and steam coal applications. In *Proceedings of the XV International Coal Preparation Congress and Exhibition*. Beijing: China National Coal Association. pp. 98–110.

Cresswell, G., and Salter, M. 2006. Design of a new era of DMS coal preparation plants for the South African Witbank Coalfield. In *Proceedings of the XV International Coal Preparation Congress and Exhibition*. Beijing: China National Coal Association. pp. 89–110.

Gray, D. 2007. The Dawson Project. In *Proceedings of Coal Prep 2007*, Lexington, KY. Stamford, CT: Penton Media. pp. 139–150.

Henry, C. 2007. Fine coal recovery plant installation at the Smith Branch impoundment—Pinnacle Mining Company. In *Proceedings of Coal Prep 2007*, Lexington, KY. Stamford, CT: Penton Media. pp. 125–138.

Hogg, J. 2007. A Canadian view. In *Proceedings of Coal Prep 2007*, Lexington, KY. Stamford, CT: Penton Media. pp. 199–205.

Hollis, R. 2007. Blue Diamond Coal Company's Leatherwood coal preparation plant undergoes major circuit modifications. In *Proceedings of Coal Prep 2007*, Lexington, KY. Stamford, CT: Penton Media. pp. 95–112.

Honaker, R., Saracoglu, M., Thompson, E., Bratton, R., Luttrell, G., and Richardson, V. 2007. Dry coal cleaning using the FGX separator. In *Proceedings of Coal Prep 2007*, Lexington, KY. Stamford, CT: Penton Media. pp. 61–76.

Luttrell, G., Honaker, R., Bethell, P., and Stanley, F. 2003. Operating guidelines for coal spiral circuits. In *Proceedings of Coal Prep AGG 2003*, Lexington, KY. Stamford, CT: Penton Media. pp. 69–78.

Metzler, S. 2007. Robinson Run: New coal preparation plant. In *Proceedings of Coal Prep 2007*, Lexington, KY. Stamford, CT: Penton Media. pp. 179–196.

Stone, R.G., Brake, I.R., and Eldridge, G. 1995. Development of a new column flotation circuit for Peak Downs coal preparation plant. In *Proceedings of the Seventh Australian Coal Preparation Conference*. Edited by J. Smitham. Queensland, Australia: Australian Coal Preparation Society. pp. 145–146.

Yanchak, D., and Metzler, S.P. 2006. Dry screening at small sizes using elliptical roller screens. In *Proceedings of Coal Prep AGG 2006*. Lexington, KY. Stamford, CT: Penton Media. pp. 37–40.

Future Challenges in Coal Preparation Plant Design and Operation

Van L. Davis Jr.

ABSTRACT

The coal industry is facing several current and future challenges that must be considered in the design and operation of coal preparation facilities. These challenges dictate that we constantly strive to improve plant operations to deal with safety concerns, demand, product quality, diminishing reserve quality, and regulatory issues. Specifically, significant problems with the capabilities and availability of plant labor, high refuse loading caused by a lower yield reserve base, fine particle dewatering, environmental problems associated with refuse disposal, and heightened monitoring and control of safety in the workplace currently are or soon will be faced by operators desiring to maximize plant efficiencies and reduce operating costs.

This chapter is an effort to provide a discussion and summary of the major challenges faced by the coal processing industry, to present a case for action, and to initiate the proper research and administrative responses. In addition, some promising recent activities in the areas of yield maximization, cost reduction, and plant efficiency improvement are discussed.

MAJOR CHALLENGES

Labor and Supervision

Coal preparation facilities undoubtedly will face labor shortages in the coming years. As shown in Table 1, the average age of a U.S. coal miner is 50 years, and this worker has a median experience of 20 years (NMA 2005). Obviously, the industry will face a significant influx of new workers as the existing staff begins to retire. The statistical data shown in Table 1 are representative of the average coal miner, not the average preparation plant operator. The preparation

TABLE 1 Profile of the U.S. coal miner

Age (mean):	50 years
Age (median):	46 years
Education*:	
High school diploma	54%
Vocational school diploma	8%
Some college	10%
College degree	5%
Work experience (median):	20 years

Source: NMA 2005.
*Data from a 1986 survey (latest data available).

plant operators at many facilities are somewhat older and more experienced than their peers who work in the mines. Therefore, it is of utmost importance that the industry immediately begin to train replacements to retain some measure of proficiency as these more experienced operators leave the workforce. Particular attention should also be paid to the training of supervisory staff to ensure that technical competency is maintained and enhanced in the next generation of plant supervisory personnel.

The implementation of enhanced programs such as behavioral-based safety should be considered as the industry begins to orient new operators to preparation facilities. Programs such as these should be implemented as soon as is feasible because inexperience with operating procedures is also evident in hazard avoidance and safe work practices.

Health and Safety

Workplace environment monitoring has received heightened attention by regulatory agencies in recent years. Specifically, compliance with ever more rigid standards governing exposure to noise is evident, and similar requirements in other areas of the workplace environment are almost certainly imminent. For instance, current noise level exposure standards dictate that an employee can never be exposed to noise levels exceeding 115 dBA, as determined without adjustment for the use of hearing protection equipment. Furthermore, no employee may be exposed to noise levels greater than an 8-hour time-weighted average sound level of 90 dBA. If this permissible exposure level is reached, a program of engineering and administrative controls must be implemented (MSHA 1999). Obviously, it would be of great benefit to maintain permissible exposure levels through engineering controls and avoid administrative controls because the latter typically involves limiting the employee's exposure by removal from the environment. In turn, this limits productivity. In addition,

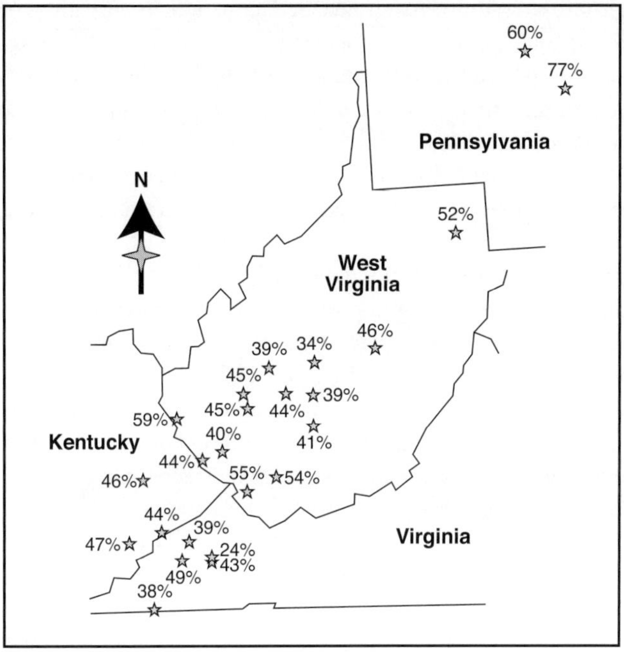

FIGURE 1 Typical preparation plant yields

hearing conservation programs involving extensive monitoring and record-keeping are required if an employee's exposure to noise levels is determined to exceed an action level of 85 dBA.

In addition to modifying existing facilities when designing and operating preparation plants of the future, the industry must focus on the engineering and construction of facilities and equipment that minimize exposure to the encumbrances of this type of regulation. Therefore, it is advisable for equipment manufacturers to design and manufacture their products to minimize noise levels. In addition, research must be performed to determine the feasibility of lowering the noise levels of existing equipment through engineering controls. The need also exists to maintain awareness of regulations during the design of new coal preparation facilities and possibly assist in the mitigation of the problem through isolation and dampening procedures.

Increased Refuse Loading

Because low-reject reserves have decreased over time, yields achieved by coal preparation facilities have declined. Typical Appalachian coal preparation plant yields are illustrated in Figure 1 to highlight current and future issues with high-refuse loading. As shown in Figure 1, it is not uncommon for preparation plants to operate at yield values in the 30%–40% range. Lower plant yield values have led to the handling of higher total raw feed tonnages for a given level of final product, increased handling and disposal of refuse caused by lower plant yields, and reduced organic efficiency in some unit operations caused by higher refuse loading. In the design of preparation plants that will be processing low-yield feed material, particular attention must be paid to refuse handling unit operations, the costs associated with refuse disposal, and the design and permitting of refusal disposal areas.

A major factor in determining an operation's competitiveness in the marketplace is cost associated with plant organic efficiency. High refuse loading to plant unit operations can reduce circuit organic efficiency, thereby increasing operating cost and decreasing yield. For example, excessively high refuse loading in dense-medium vessels increases the likelihood of coal losses to refuse by entrapment. Recoverable particles can easily be trapped by refuse because of the increased mass flow of particles to the reject stream. In addition, adverse affects can be found in dense-medium cyclone circuits, particularly those processing a wide size distribution of feed material, when refuse loading becomes abnormally high for the unit's design. When mass flow of refuse to a cyclone's apex becomes high, it is fairly common for finer lower-specific-gravity material to be trapped in the refuse flow. This lowers circuit organic efficiency and yield. In the future, therefore, it is imperative that research be performed on equipment design that will minimize the negative impact of high refuse loading. In addition, during the design phase of new facilities and circuit modifications, particular attention should be paid to the range of size distributions fed to each unit operation. A reduction in plant organic efficiency can quickly offset the capital savings realized by feeding a wide range of particle sizes to fewer unit operations.

Environmental Issues Associated with Refuse Disposal

A convenient and cost-effective method for disposing of fine refuse from coal preparation plants is underground injection. The use of this technique has many advantages when compared with other methods of fine refuse disposal. The onerous permitting procedures and long lead times associated with fine refuse impoundments are avoided. The construction of fine refuse impoundments is becoming increasingly more difficult and time consuming, especially because recent publications have advocated the study of alternatives to reduce or eliminate the need for impoundments (Committee on Coal Waste Impoundments et al. 2002). The high unit costs associated with the operation of belt presses or vacuum filters are also circumvented. However, the presence of hydrocarbons and other chemicals deemed unacceptable by regulatory agencies in fine refuse and its disposal has become a challenge to current coal

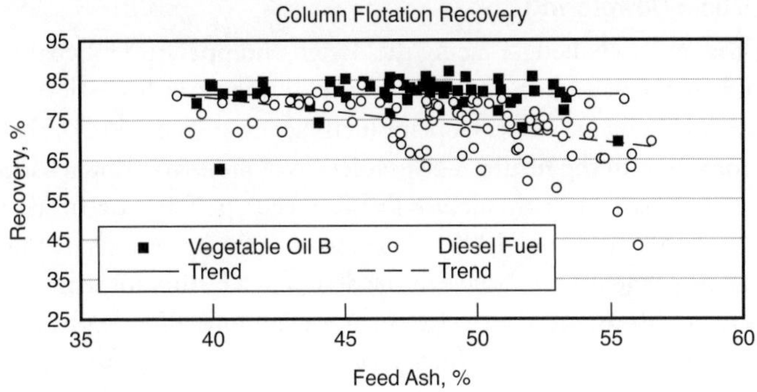

Courtesy of Freedom Industries.

FIGURE 2 Alternate collector versus diesel fuel comparison

processing activities where underground injection is a viable alternative. An extensive list of acceptable and unacceptable chemicals for underground injection has been compiled by the Division of Water and Waste Management in the state of West Virginia (Pettigrew 2007). Unfortunately, many common flotation reagents, flocculants, and clarification aids are banned from use in underground injection operations. Most importantly, the use of diesel fuel as a flotation collector is not permitted when fine refuse is injected underground into old mine works. Fortunately, alternative flotation collectors are available that are acceptable for use in underground injection. As illustrated in Figure 2, the use of some of these environmentally friendly collecting agents has demonstrated equivalent and, in some cases, better flotation performance when compared with diesel fuel (Skiles and Kennedy 2003).

A future challenge associated with fine refusal disposal is the potential for applying the same restrictions to flotation and thickening chemicals in operations using impoundments. Studies have been conducted by those (Patel and Schreiber 2001) in other mining industries to address environmental concerns surrounding chemical migration, in particular hydrocarbons, into groundwater supplies. This study was conducted in the phosphate industry in Florida, where the use of impounded areas is the only means of fine refuse disposal. With greater restrictions on the likely use of flotation and thickening chemicals, the proactive approach is to conduct studies and make the final determination that a problem with the migration of harmful substances into groundwater actually does exist and, if necessary, initiate the use of alternatives. The use of additional alternative reagents and the enhancement of existing acceptable chemicals should also be investigated.

Fine Particle Dewatering

One of the major challenges facing the design and operation of coal preparation plants is the dewatering of fine particles. Dealing with the dewatering of fine particles is a major source of plant inefficiency and cost and probably will be a factor well into the future. The prediction of moisture content during the design phase has met with success in the recent past by use of algorithms defined by others (Arnold 1999). The use of techniques such as that referenced can be particularly effective in the design of a facility for which no existing process performance data exist. Such a technique can also be useful for demonstrating the impact on final product moisture that can be made by changes in processing circuitry or equipment. However, the fundamental problem with excessive moisture content in fine coal streams, particularly flotation concentrates, still exists because of the lack of technology for efficient and cost-effective removal.

Plant yield is maximized when the last particle recovered from each circuit is of the same quality (Luttrell et al. 2003). In other words, yield is maximized by maintaining constant incremental quality. This is typically understood as maintaining constant incremental ash content, which is correct if moisture content is neglected. This principle generally holds in most metallurgical coal plants where thermal drying unit operations exist. In steam coal plants, however, where thermal drying operations are not typically found, moisture must be accounted for and yield is maximized through constant incremental inert content in each circuit. Because every unit of water recovered in the product stream is 100% inert, the displacement of this water would allow the recovery of more that one unit of weight of middlings particles by increasing operating specific gravities. Typically, the removal of 1 ton of water allows the recovery of 3 tons of middlings particles (Luttrell et al. 2003). Table 2 contains a typical column flotation concentrate size distribution from a steam coal plant in West Virginia. This material is currently combined with product from a spiral and water-only cyclone circuit and dewatered in screen-bowl centrifuges. Typically, approximately 50% of the minus-325-mesh material is lost to the effluent in the centrifuge circuit. As shown in the table, the amount of

TABLE 2 Column flotation concentrate size and quality

Screen Size	% Weight	% Ash
+28 mesh	0.00	7.27
28M × 60M	0.70	5.01
60M × 100M	3.86	5.64
100M × 200M	18.87	6.60
200M × 325M	16.31	8.98
325M × 0	60.25	14.46
Total	100.00	11.68

material in the minus-325-mesh fraction is approximately 60% of the product, and the product ash content is acceptable and yields a British-thermal-units-per-pound value in excess of 13,000 on a dry basis. Unfortunately, this material in its "as received" form is more than 50% moisture; therefore, its incremental inert content is more than 57%. When this stream is compared with the incremental quality of middlings particles from coarser circuits at approximately 35% incremental ash and 5% moisture, the incremental inert valuc of the middlings particles is slightly more than 38%. In other words, the removal of 1 ton of high-moisture fine material allows the recovery of 1.5 tons of middlings, thereby increasing plant yield. This effect is the main reason for the rise in popularity of deslime cyclone circuits ahead of flotation. However, if a technique is developed that could lower the moisture of this fraction significantly and result in an incremental inert content similar to that of the coarser middlings fractions, even larger plant yield gains could be realized by recovering more of the particles in both streams. As illustrated in Figure 3 (Luttrell et al. 2003), it is not uncommon for the minus-100-mesh fraction of a plant's product to contain more than 30% of its total product moisture. These issues

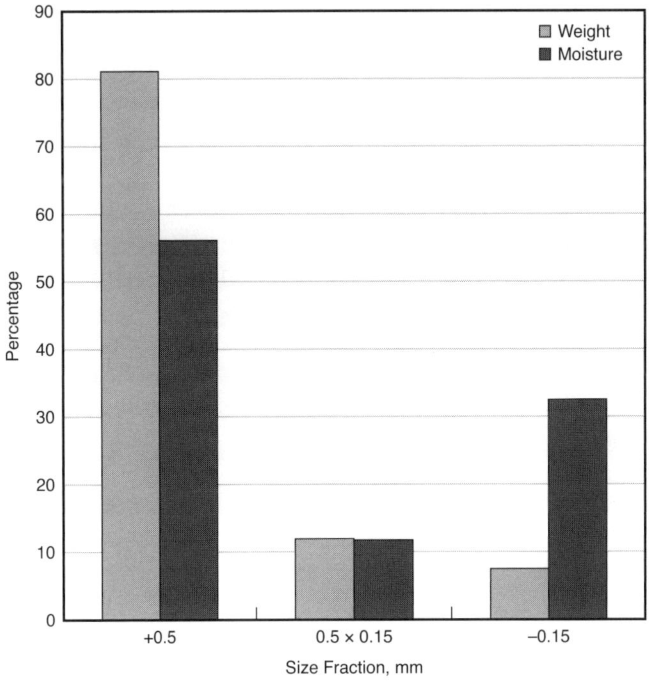

Source: Data from Luttrell et al. 2003.

FIGURE 3 Percentage of dry weight and total moisture by size fraction

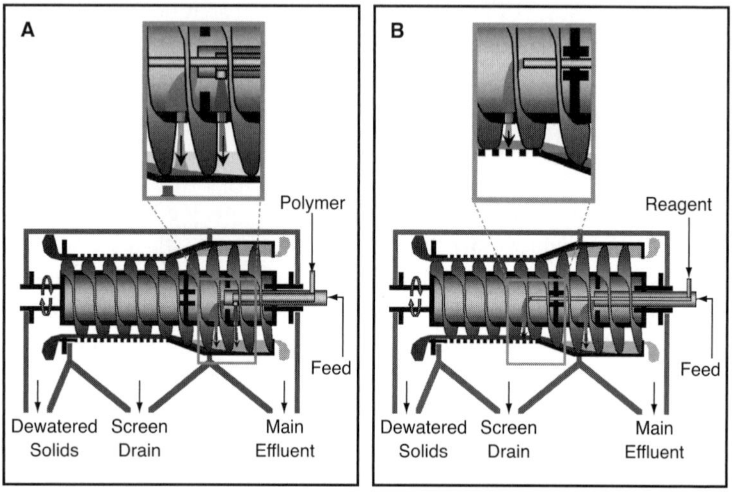

Source: Luttrell et al. 2006.

FIGURE 4 Schematic showing the injection tubes for (a) adding flocculant directly into dilute low-solids pool in the bowl and (b) adding surface tension modifiers to partially dewatered cake on the screen section

present a valid case for aggressively pursuing research in the area of fine particle dewatering improvement.

Another area for future improvement in plant design and operation would be to improve the recovery of the fine particles currently lost in centrifugation and enhance dewatering technology to improve plant yield. Work performed by Luttrell et al. (2006) shows promise in this area and merits further investigation on an industrial scale. In this work, the researchers incorporated the use of injection tube technology on screen-bowl centrifuges for industrial-scale trials of flocculant injection and pilot-scale trials of the injection of surface tension modifiers for moisture reduction. The methods for chemical injection are illustrated in Figure 4. This flocculant injection technique resulted in the recovery of approximately 93% of the material previously lost to the centrifuge effluent. However, the recovery of these particles resulted in an increase in the centrifuge product moisture of approximately 10%. The pilot-scale test work associated with the injection of surface tension modifiers showed promise in the area of moisture reduction, but effectiveness on an industrial scale and the associated cost of the reagents must be investigated in more detail to determine economic feasibility.

SUMMARY

It is evident that many challenges, a few of which have been presented here, are facing those involved with the future design and operation of coal preparation plants. These problems should be addressed directly and aggressively to ensure that the industry remains competitive in the areas of employee health and safety, product quality, proportion of coal reserves converted to salable product, environmental stewardship, and operating costs. Many of the areas of research presented here show promise for the future, and possibly the present. At the same time, we need to evaluate the current regulatory climate and address restrictions that are not based on scientific fact. Accordingly, we must educate our peers in government and society as to the basis of our arguments and present ourselves as a needed industry for the country's economic prosperity. As an industry, we need to strive for constant improvement and continue to evaluate emerging technologies and put them to use for the economic benefit of our industry.

REFERENCES

Arnold, B.J. 1999. Simulation of dewatering devices for predicting the moisture content of coals. *Coal Prep.* 20:35–54.

Committee on Coal Waste Impoundments, Committee on Earth Resources, Board on Earth Sciences and Resources, National Research Council. 2002. *Coal Waste Impoundments: Risks, Responses, and Alternatives.* Washington, DC: National Academy Press.

Luttrell, G.H., Burchett, R.T., and Yoon, R.-H. 2003. Economic benefits of improved moisture reduction in coal preparation plants. In *Proceedings of the 20th Annual International Pittsburgh Coal Conference*, Pittsburgh, PA, September 15–19. Preprint 16-1. Pittsburgh: Coal Conference.

Luttrell, G.H., Burchett, R.T., McGough, K., and Stanley, F.S. 2006. Improving screen-bowl centrifuge performance. In *Proceedings of the SME Annual Meeting and Exhibit*. SME Preprint No. 06-010. Littleton, CO: SME.

MSHA (Mine Safety and Health Administration). 1999. Occupational noise exposure. Title 30, Code of Federal Regulations, Part 62. Arlington, VA: MSHA.

NMA (National Mining Association). 2005. Profile of the U.S. coal miner. www.nma.org/pdf/c_profile.pdf. Accessed July 2007.

Patel, S.K., and Schreiber, A.E. 2001. *Fate and Consequences to the Environment of Reagents Associated with Rock Phosphate Processing*. Bartow, FL: Florida Institute of Phosphate Research.

Pettigrew, P.L. 2007. Underground injection control: Mining. List of acceptable, unacceptable, and not rated, groundwater/underground injection control. Charleston, WV: Department of Environmental Protection, Division of Water and Waste Management.

Skiles, K.D., and Kennedy, D.L. 2003. Search for the next generation of coal flotation collectors. In *Proceedings of the 20th Annual International Coal Preparation Exhibition and Conference*, Lexington, KY, April 29–May 1. Greenwood Village, CO: Primedia Business Exhibitions. pp. 187–204.

PART 2

What We're Doing

Coal Crushing Considerations

S.W. Eck

ABSTRACT

As coal preparation plant circuit design has evolved, the need to crush raw coal before feeding has also evolved. In modern circuits, crushing coal to –2 in. is the latest trend. Crushing raw coal to this size allows more complete separation of the coal from the rock and mineral matter, and maximizes coal recovery. The raw coal crusher must be carefully selected to minimize creation of additional fines because fine coal cleaning is more difficult and more costly. The modern low-speed sizer was developed for this purpose and is rapidly becoming the preferred crusher. Sizers are also being installed underground to precrush raw coal for more efficient belt haulage system operation. Future deep-cleaning plants designed to prepare coal for gasification and liquefaction facilities will also use sizers as the crushers of choice.

TRADITIONAL COAL CRUSHING REQUIREMENTS

Coal preparation has been a practice ever since the early days of coal mining. In its simplest form, coal preparation started by handpicking rock and other debris that were a natural part of the mined coal. Hand-mining techniques allowed the selective separation of rock at the face, surface stockpile, or tipple. Because the coal was hand mined at a low rate, hand separation was practical. With the advent of early mechanized mining and coal preparation plants, handpicking was also used to remove rock and other debris ahead of screening and washing circuits before shipment to customers. As the rate of coal flow and capacity of the preparation plant increased, handpicking became impractical and prohibitively labor intensive.

Early coal preparation circuits using selective screening, jigging, and washing tables became the acceptable industry design. To prepare the raw coal feed to these devices, typically coal crushing to –6 in. was needed. With the introduction of artificial gravity separation devices such as Chance cones and magnetite-based dense-medium slurry vessels, precrushing to –6 in. was also necessary. Single-roll and double-roll crushers were usually used for this crushing duty (Figures 1 and 2). When the end use of the clean, washed coal was for the industrial and home-heating markets, the resultant clean coal was screened into various well-defined size components. When the end use of the clean, washed coal was for electric utility boilers, final crushing to –2 in. was needed. Crushing this clean coal remains an operation best suited for double-roll and ring-hammermill-type crushers (Figure 3).

FIGURE 1 Typical single-roll crusher

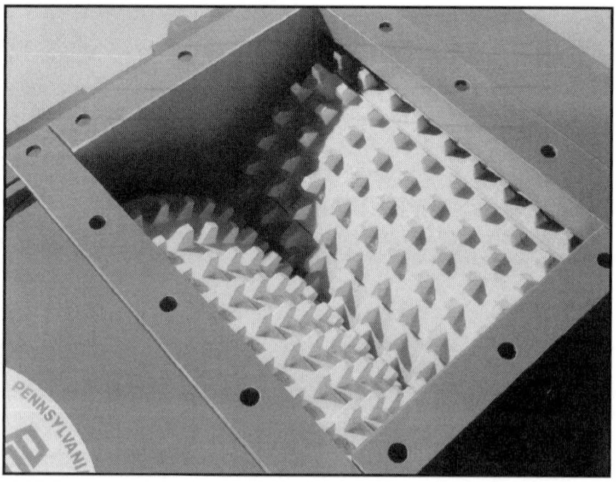

FIGURE 2 Typical double-roll crusher

COAL PREPARATION STARTS AT THE FACE

With all coal mining systems, coal preparation starts at the face. Whichever mining system used—hand mining and loading, continuous mining, and so forth—the initial coal quality and coal recovery efficiency start at the mining face. The lower the amount of extraneous material kept out of the raw coal transported to the preparation plant, the higher the coal recovery percentage at the plant and the lowest possible overall mining costs.

Given the natural inclusion of discrete mineral particles, mineral interbands in the coal seam, and multiple thin coal seams separated by mineral seams, it is not often possible to eliminate these minerals at the face. In areas of soft clay bottoms or thin, poorly laminated shale roof structure, which cannot be supported successfully, various amounts of rock are mined along with the coal. Some of this rock, particularly slabby roof material, is transported to the preparation plant to be separated. It is the job of the raw coal crusher to reduce the size of this rock material, and the coal particles, to a maximum size that can be accepted by the coal separation equipment. Dense-medium baths, dense-medium cyclones, and froth flotation have been used for more than 40 years as high-efficiency separation devices for the traditional coarse, medium, and fine size fractions of the coal size distribution fed to the preparation plant. Each of these separation devices has maximum and minimum particle size parameters. Generally, the coarser the particle size, the lower the cost of separating coal from rock. Therefore, to minimize preparation costs and

FIGURE 3 Typical ring-hammermill crusher

maximize efficiency and recovery percentages, it is necessary to minimize the amount of additional fines created by the raw coal crusher.

Immediately after the coal mining face, the raw coal must be transported to the surface for processing. This transportation system, either underground or on the surface, typically is a belt conveyor. Conveyor systems usually have numerous transfer points, and storage and reclaim systems. Large pieces of coal and large slabby pieces of rock can cause numerous problems with the transport system. Large pieces can become jammed in the transfer and storage reclaim points, causing systemwide pluggages. These pluggages must be eliminated, typically by hand labor, and any coal spills cleaned up. Large pieces can also knock the belt conveyor from its centralized position on the troughing idlers, shutting down the conveyor system because of misalignment; drop onto the lower belt at transfer points, punching holes in the belt; or lodge in a transfer point in such a way as to longitudinally cut the receiving belt as it passes. During this unscheduled downtime, coal production from the mine usually ceases, decreasing overall productivity and causing high belt repair or replacement expenses.

Crushing directly by the mining face in both underground and surface mines can minimize or eliminate belt haulage problems associated with large particles in the raw coal stream. Such crushing systems are being installed or retrofitted in both types of mines. Given the trend to larger-capacity mines, these raw coal crushers will also be of larger capacity. It is not unusual to require the raw coal crusher to process 6,000–8,000 tph (Figure 4).

Another consideration is the coal size distribution after mining and crushing. In general, the smaller the coal and rock particles, the more mineral matter is liberated from the coal and the more difficult and more costly it is to separate the two. The need to crush the coal to achieve the most efficient separation of the coal from the mineral matter and the highest recovery of coal must be balanced with the need to reduce the process costs of fine particle cleaning circuits. Also, the smaller the overall particle size distribution, the more surface area obtained and the more surface moisture remaining after washing and mechanical drying. In mining areas that have naturally occurring small particle size distributions, typically coals with a high Hardgrove grindability index value, thermal drying has been used to reduce overall surface moisture to acceptable levels. Thermal drying contributes to higher overall mining costs compared with coals that can be mechanically dried and meet customer needs. It becomes apparent that the coal preparation process is greatly influenced by and begins with the mining system design. System design, along with the coal's intended end use, determines the coal preparation flowsheet design.

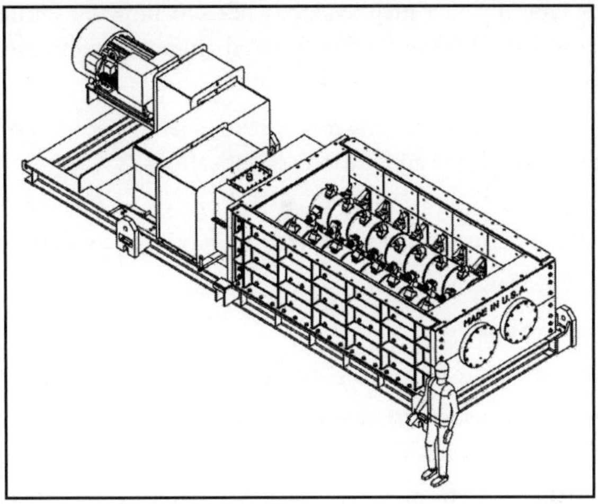

FIGURE 4 Typical larger size crusher for underground installations (7,000 tph raw coal, 12 in. × 0 product)

LOW-SPEED SIZERS

In the quest for more efficient raw coal crushing, which does not generate excessive fines, low-speed sizers were developed. Advantages of the low-speed sizer in comparison to the traditional single-roll, double-roll, or hammermill-type crushers are as follows:

- Sizers have very strong steel crusher box frames to contain all crushing forces in the box without transmitting these forces to the support structure.
- Sizers use shear forces to crack all oversize pieces in the feed stream.
- Sizers allow all product size particles to flow through the unit without being crushed. This greatly reduces needed power and minimizes generation of additional fines.
- Sizers operate at low rotor shaft speeds to minimize wear and maximize torque at the rotor tooth–coal contact point. This also minimizes the generation of additional fines.
- Sizers have easily accessible, replaceable wear components.
- Sizers have low height dimensions.
- Sizers typically use simple grease-lubricated rotor shaft bearings.

- Sizers typically have maintenance wheels to allow the entire sizer to be easily rolled out from under the feed chute, allowing easy access for maintenance.
- Sizers are available in many model sizes to allow selection of the proper machine to suit the application's specifications.
- Sizers are available as center sizers and side sizers. Center sizers usually are primary crushers, and side sizers usually are secondary sizers.
- Sizers have low drive-motor power needs.
- Sizers have low noise emissions.
- Sizers can handle wet, sticky feed material and a wide variety of coals.
- Sizers can be installed as a two-stage crushing system with the primary sizer located directly above the secondary sizer.
- Sizers can be fed from either side or from over the nondrive end.
- Sizers can be fed by any type of feeder or by direct belt conveyor discharge.
- Sizers typically have a fixed roll shaft rotational speed and a simple operating mode. The drive motor is either running or not.
- In sizers with a single-motor drive using gears to link the rotor shafts, the rotor shaft teeth are always in alignment. Two-motor drives operate a majority of the time with the teeth out of alignment. Misaligned teeth allow large, flat, slabby oversize pieces to pass through the sizer without being crushed, potentially damaging downstream process equipment or causing transfer point blockage or conveyor belt damage. Sizers such as the Pennsylvania Crusher Corporation's Mountaineer sizer with linked rotor shafts do not allow passage of large, flat, slabby oversize pieces.

It becomes apparent that a low-speed sizer is an excellent crusher for reduction of raw and clean coal feeds where generation of fines is a requirement.

Sizers can also be used as raw coal crushers for direct shipment installations. These are storage and rail loading systems that do not have an associated coal preparation plant. The as-mined raw coal quality is high enough to be crushed to –2 in. and loaded directly into railcars or trucks for shipment to utility generating plants (Figures 5–8).

PRESENT AND FUTURE CONSIDERATIONS

The present and future trends for more efficient and higher-capacity coal preparation plants, in terms of rock and mineral matter separation and coal organic recovery, will continue. It is desired that this increase in capacity and efficiency be accomplished at the lowest possible cost. Changes can be seen in flowsheet design and separation equipment performance to accomplish this

FIGURE 5 Sizer crushing 800 tph raw coal (2 in. × 0 product) at direct-ship facility in eastern Kentucky

FIGURE 6 Sizer crushing 600 tph raw coal (6 in. × 0 product) at preparation plant in eastern Kentucky

higher separation and recovery efficiency. Most notably, the introduction of fine particle dense-medium cyclones and the lowering of plant feed coal size distribution to −2 in. is becoming prevalent. Lowering the particle size distribution to −2 in. typically liberates a greater portion of the coal from the associated rock and mineral matter than when the coal is crushed to −6 in. Offsetting this increase in separation efficiency and recovery is the higher

FIGURE 7 Sizer crushing 2,200 tph raw coal (6 in. × 0 product) at preparation plant in eastern Kentucky

FIGURE 8 Sizer crushing 400 tph raw coal (4 in. × 0 product) at preparation plant in People's Republic of China

percentage of fines contained in the coal feed stream and the need for additional capacity in the more expensive fine coal cleaning circuits of the preparation plant. Fine coal cleaning circuits—usually dense-medium cyclones, spirals, and froth flotation—are becoming the accepted norm for plant flowsheets. This dictates the crushing of all incoming run-of-mine coal to –2 in. or –4 in. when the latest larger-diameter dense-medium cyclones are used. At the same time, it is still imperative to minimize the amount of additional fines in the plant feed. Relying on the older-design raw coal crushers for higher plant feed rates at smaller output sizes will probably lead to failure. New raw coal crusher designs, such as low-speed sizers, are becoming the only alternative to meet these crushing needs.

Coal preparation plants of the future must also be designed to prepare feed for coal gasification and coal liquefaction process systems. These new systems will produce the fuels of tomorrow from abundant coal reserves in an acceptable, environmentally friendly manner. The same trends of crushing raw coal for preparation plant feeds apply to these future process systems (i.e., the generation of minimum fines during crushing). The coal fed to gasification and liquefaction process plants usually is in the form of a water slurry. To pump coal–water slurries into the gasification and liquefaction retorts requires a slurry of high percentage solids and low viscosity. Maintaining this ratio of high coal solids to water in the slurry is not possible with excessive fines. The generation of low fines in the coal slurry starts with the first raw coal crusher and proceeds through all stages of coal crushing and coal preparation. Selection of a coal crusher that generates excessive fines anywhere in the coal preparation process reduces the efficiency of the gasification or liquefaction process.

Is There Anything New in Coarse or Intermediate Coal Cleaning?

Dieter Ziaja and Giancarlo F. Yannoulis

ABSTRACT

Over the years, coal cleaning technologies have evolved to meet the demands of the industry. This chapter describes the development and current features of the unit process equipment used for the cleaning of coarse and intermediate coal, specifically jigs and dense-medium separators.

INTRODUCTION

Is there anything new in coarse or intermediate coal cleaning? The answer is easy: Yes, there is.

Technologies for coarse coal separation have been available as far back as the 19th century. Over the years, advances in electronics, instrumentation, material science, and so on have ensured that these technologies will be with us for a long time.

In today's competitive world of coal cleaning, cost-effective beneficiation is of the utmost importance. In general, separation of coal at coarse sizes or the removal of a portion of refuse from uncrushed raw coal offers great advantages, including the following:

- Inexpensive production of steam coal without the need for magnetite
- Decreased power needed for size reduction to product size
- Decreased generation of high-ash fines
- Decreased unit surface area and hence lower product moisture
- Increased homogeneity of the coal feed to subsequent processes
- Decreased wear of downstream preparation equipment
- Improved use of preparation plant capacity

This chapter describes the developments and current features of unit process equipment for the following applications:

- Primary treatment of run-of-mine (ROM) coal with movable sieve bed jigs (Romjig)
- Separation of coarse and intermediate coal by under-bed air-pulsated jigs (Batac)
- Separation of coarse coal at low cut points by dense-medium separators (Teska)
- Separation of coarse coal by large-diameter dense-medium cyclones

PRIMARY TREATMENT OF ROM COAL WITH MOVABLE SIEVE BED JIGS (ROMJIG)

Jig separations use bed pulsation (or jigging) to achieve bed segregation by density. This is a cost-effective way to remove extraneous rock from ROM or raw coal. Where the first separation objective is a virtually coal-free refuse stream, this is commonly called de-stoning. Two jigging technologies have been developed to suit today's coal industry needs.

The first technology is the Romjig, previously described in works by Wesp et al. (1989), Ziaja and North (1998), and Sanders et al. (2000, 2002). The second technology is the adaptation of the Batac jig for de-stoning applications, previously described in works by Ziaja and North (1998) and Sanders et al. (2000, 2002).

This chapter presents and reviews recent applications using both of these technologies in the South African coal industry. The primary focus is the role of jigging in the overall process scheme, installation costs, and the design approach needed to achieve the best performance.

Design and Operation of the Romjig

Figure 1 (Sanders and Ziaja 2003) shows the design features and operating principles of the Romjig. The process is designed to treat raw coal within a particle range from 350 mm to 40 mm. The 350-mm top size is the maximum, and a lower top size is equally acceptable.

Based on processing and handling considerations, the minimum particle size limit is 40 mm. Presized raw coal is fed into the Romjig onto an inclined movable screen submerged in water in the main vessel, which contains about 30 m^3 of water. The perforated screen is pivoted at its discharge end.

The feed end of the screen is lifted and then allowed to drop by gravity to create the jigging action. The availability of high-capability hydraulic power is one of the key factors in the development of this process.

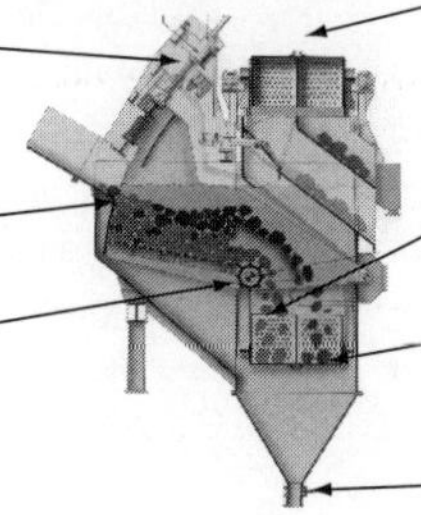

Source: Sanders and Ziaja 2003.

FIGURE 1 Design and construction of Romjig 3 Generation

The lifting cycle is repeated 38 to 43 times per minute. Under the jigging action the material separates, with the heavy particles congregating next to the screen, progressively forming a bed as the material moves along the sieve, while the lighter coal moves to the top.

Three factors ensure that the material moves horizontally from the feed end to the discharge end:

- The slope of the screen deck (pressure from the feed)
- The linear downstroke
- The circular upstroke

Products are collected in a divided discharge wheel with one compartment for the coal and one for the discard. Particles smaller than the screen panel apertures fall through into the hutch compartment. Material accumulates in the hopper at the bottom and is discharged periodically through a double-gate valve system to join with the minus-40-mm feed screen underflow material for further treatment or as final product.

To ensure that design and manufacturing costs remain competitive, the Romjig is available in only one size, with a 2.0-m-wide bed, rated at up to 350 t/h for a nominal 350–40 mm feed. Details of the construction and operational design are summarized in Table 1.

The control of the pulsation is illustrated by the simple but effective principle depicted in Figure 2. Control is achieved by monitoring the pressure exerted on the rocker arm during the upward movement of the bedplate.

This pressure is a function of the weight of material on the bedplate and is proportional to the amount of heavy material forming the bed. The pressure reading is used to regulate the rotational speed of the discharge roller and the frequency of the pulsation. At very low pressures, which indicate a thin, light bed, the roller discharge is stopped to prevent the loss of bed while the pulsation is

TABLE 1 Romjig details

Construction		Process Details	
Length	~6.0 m	Feed rate	350 t/h
Width	~6.5 m	Feed size	350–40 mm
Height	~8.0 m	Stroke amplitude	500 mm
Bed width	~2.0 m	Stroke frequency	38–43/min
Full weight	98 tonnes	Makeup water requirement	10–15 m^3/h
Installed power	110 kW	Specific energy	0.3 kWh/t
		Cut-point range	1.6–2.1 RD*

*RD = relative density.

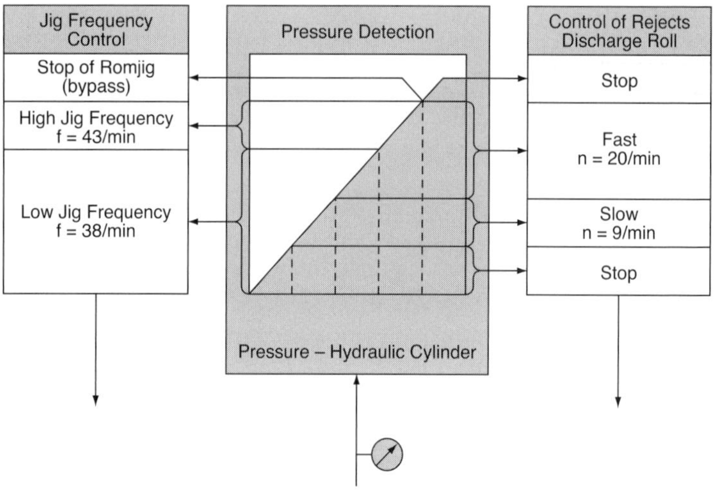

FIGURE 2 Romjig pulsation and bed control

kept at its slowest. With these conditions, the bed is retained and gradually built up to a normal working level from the incoming rejects in the feed.

At a moderate pressure, representing a lighter bed containing some heavy material, the discharge roller is operated at a slow speed while the pulsation frequency is kept to the lower limit of approximately 38 strokes/min. When the feed delivers more discard material, the pressure on the rocker arm increases. The control system reacts by increasing the discharge roller speed to extract the rejects at a higher rate. Should the discard load in the feed exceed the maximum extraction capability of the machine, the system activates a bypass facility, and pulsation stops.

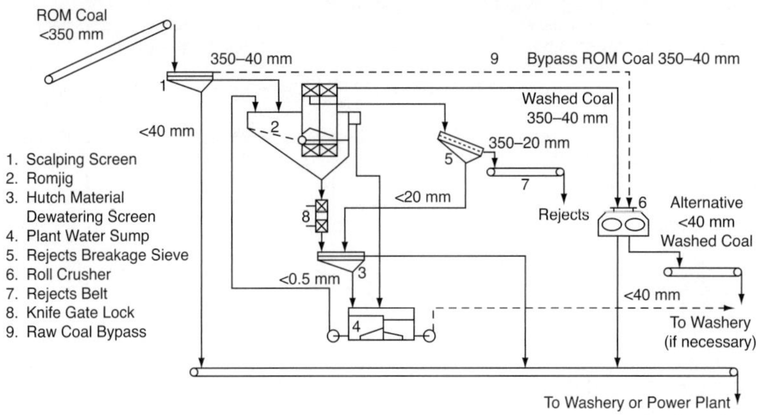

Source: Kottmann et al. 2006.

FIGURE 3 Typical Romjig flowsheet

The discharge roller also stops to prevent the loss of bed. When the feed rate increases, the stroke frequency is increased to handle the higher load. The controls can be integrated with on-line ash and size analyzers. It has been found in practice that the point of critical control is the discharge roller speed, and most operators leave the pulsation frequency fixed all the time.

A typical circuit for a Romjig plant is shown in Figure 3. Dewatering of clean coal takes place in the discharge wheel. In this application, clean coal is crushed to suit further treatment and joins the untreated minus-40-mm raw coal stream.

Rejects are additionally dewatered and sized on a screen to recover smaller particles, which tend to be coal. The rejects screen oversize is discarded. The water and undersize from the rejects screen gravitate to the hutch product screen (a high-speed vibrating screen), which also collects the periodically discharged hutch product.

The hutch product screen overflow, essentially fine coal, is added to the main coal product, while the water is collected in the main sump for reuse in the process. The process produces very small amounts of fines, so no additional water clarification system is needed.

Figure 4 illustrates the water usage of the process. Total makeup equals the surface moisture lost with the product and discard (this is low because of the small surface area of the large sizes processed), plus the small amount of water discharged with the small amount of fines from the hutch product screen. The low water consumption makes the process attractive for remote locations.

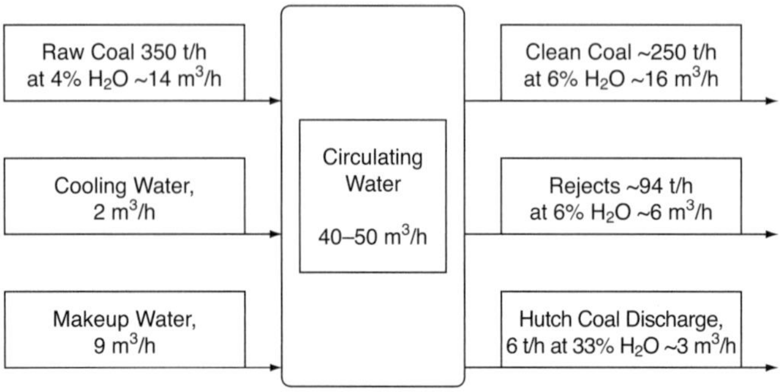

FIGURE 4 Typical Romjig water balance

Operation of Romjigs at BHP Billiton's Colliery

At BHP Billiton's Optimum colliery in South Africa, two Romjigs are in operation for de-stoning of the coarser ROM coal. The mined coal is screened at 50 mm. A scalping sieve ensures that material +350 mm is bypassed.

From the screening section, the 350–50 mm material is conveyed to the plant storage-and-feed bin, ensuring continuous operation of the Romjig plant. To minimize breakage, each Romjig is fed via a separate screen feeder located at the bin outlets. A bypass system has been installed for emergencies or when it is not necessary to use the de-stoning plant (Figure 5).

The justification for this Romjig installation stems from an increase in ROM dilution, resulting in more than 30% rock in the plant feed. It was decided to build a de-stoning plant in front of the existing dense-medium washery. The Romjig plant was built and commissioned in 2001. Design, capital, and operating data are shown in Table 2.

SEPARATION OF COARSE AND INTERMEDIATE COAL BY UNDER-BED AIR-PULSATED JIGS (BATAC)

Design and Operation of Batac Jig

A short Batac jig, which is a jig with a length of only two or three chambers, can be used to de-stone raw coal ahead of a dense-medium circuit (Figure 6). The key difference between a Batac jig for de-stoning applications and a Romjig is the size range of the material to be beneficiated. Batac jigs are used in South Africa for material with a size range from 60 to 6 mm. Thus they would typically be fed with raw coal rather than ROM coal. In general, a feed top size up to about 150 mm can be accommodated.

The jigging motion is generated in air chambers located underneath the jigging bed. Low-pressure, high-volume air from a blower is intermittently

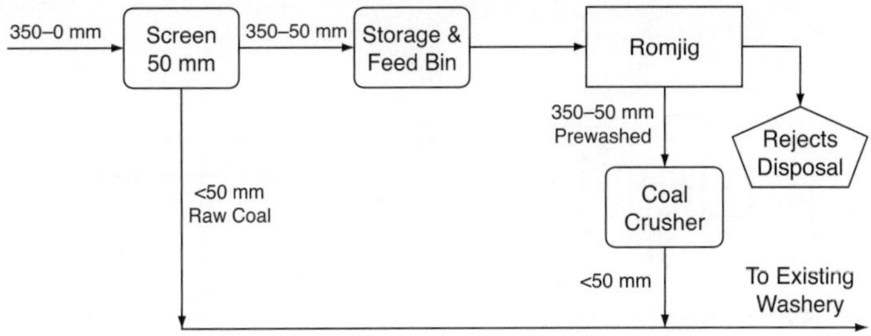

FIGURE 5 Block flowsheet of Optimum colliery operation

TABLE 2 Design, capital, and operating data for the Optimum colliery

Data	Romjig I	Romjig II
Design data		
Material	Raw coal	Raw coal
Feed size	350–60 mm	350–60 mm
Feed rate	350 t/h	350 t/h
Operating data		
Rejects in feed	>30%	>30%
Rejects in float at 2.4 t/m^3	3.8%	5.8%
Coal in sinks at 1.8 t/m^3	1.55%	1.02%
Capital costs, in U.S. dollars*		
Screening section	$350,000	
Feed bin and conveying section	$500,000	
Romjig section	$3,850,000	
Crushing and conveying section	$650,000	
Operating costs, in U.S. dollars*		
Power consumption screening section, per year	1.4 million kWh, $65,000	
Power consumption Romjig section, per year	3.0 million kWh, $140,000	
Power consumption crushing section, per year	4.6 million kWh, $210,000	
Water consumption (including nonprocess water)	30 m^3/h	
Consumables, per year	$40,000	
Labor costs, per year	$120,000	
Maintenance and spares, per year	$140,000	

*Based on year 2006 for two machines.

supplied to these air chambers and then discharged by means of an electronically controlled plate valve system. The frequency and profile of the resulting stroke can be readily modified with the Batac jig programmable logic controller (PLC).

The stroke is imparted to the water inside the jig as a function of the pressure change generated inside the air chambers. Makeup water is added at the lowest point of every jigging chamber to intensify the upward current and to restrain the downward current. The pulsating motion of the water stratifies the feed material according to its density.

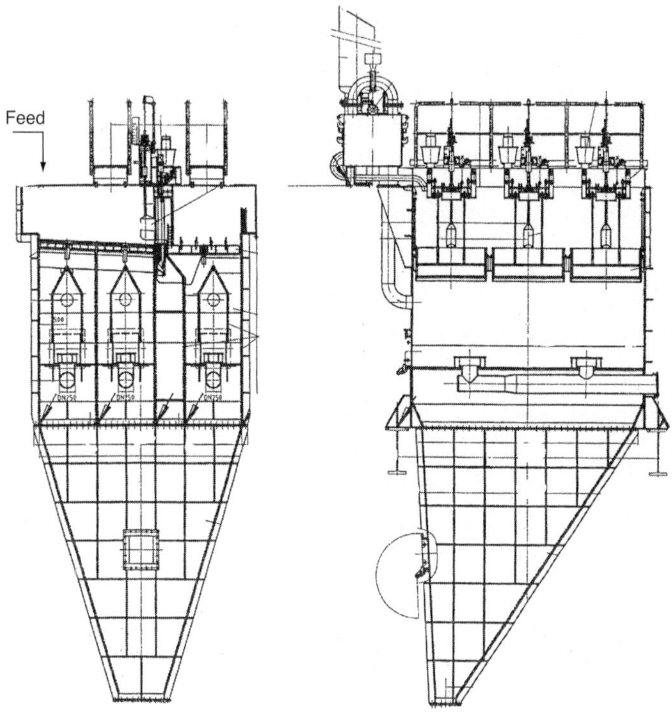

FIGURE 6 Batac jig principle

The magnitude and shape of the stroke used in the jigging process are essential for successful separation. Therefore, the design of a pulse generator that can guarantee a controlled supply of compressed air is crucial for the process (Figure 7).

As mentioned earlier, the flow of jigging air is controlled electronically and can be set individually for each jigging chamber. A common pulsation frequency (ranging from 40 to 120 strokes/min) for all jigging chambers is set in the jig's PLC. This is a conventional unit available from many leading PLC suppliers.

The thickness of the rejects layer of the stratified material is scanned according to specific gravity by a sensor float connected to an appropriate measuring and recording system via ultrasonic velocity probes (Figure 8). The basic float setting is performed through addition or removal of weight to the float. The discharge system gate is actuated by a hydraulic unit, which in turn is controlled by a proportional-integral-derivative circuit integrated within the jig PLC. Feedback from the ultrasonic sensor ensures appropriate opening and

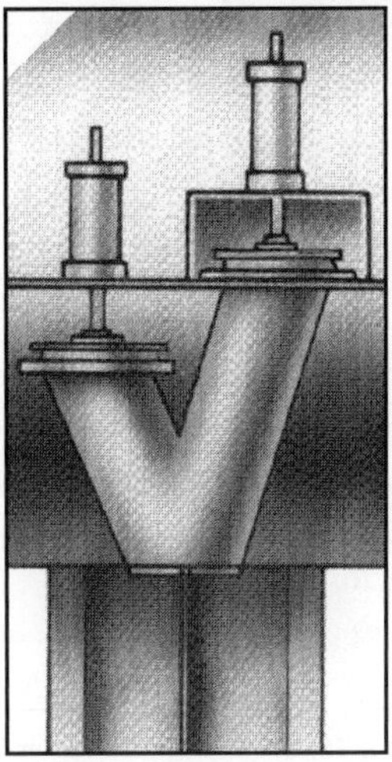

FIGURE 7 Batac jig pulse unit

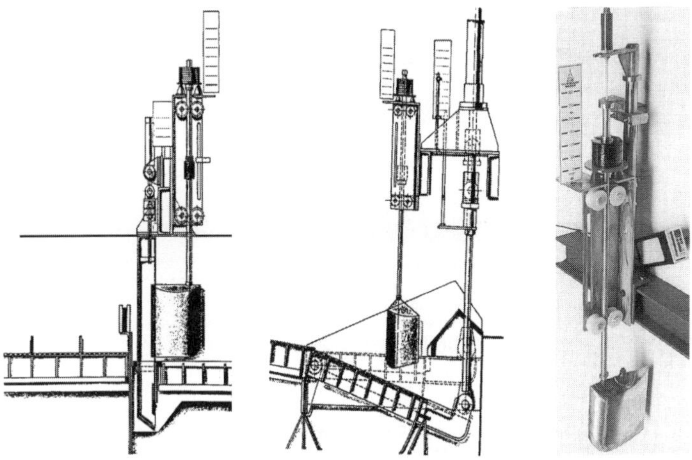

Fine Coal Discharge Device
(Feed Size: <80–0.5 mm)

Coarse Coal Discharge Device
(Feed Size: <150–10 mm)

FIGURE 8 Batac jig discharge system

closing of the gate to ensure correct withdrawal of reject material. Jigs with large widths are equipped with independently operating discharge devices, complete with separate sensors and hydraulic units.

Operation of Batac Middle Coal Jig

De-stoning is also the reason why a jig was selected for the Leeuwpan plant in South Africa. The concept differs from the Optimum de-stoning plant because of the different size range. At Leeuwpan, the complete plant feed material (550 t/h) is crushed to pass 60 mm using a roll crusher and then screened at 8 mm (slotted) to produce –6 mm and 60–6 mm size fractions (Figure 9).

The 60–6 mm material is treated in a 4-m-wide, three-compartment Batac jig for de-stoning. The minus-6-mm material leaves the plant untreated, optionally as product. A blending device allows the plant to blend the 60–6 mm washed coal and the minus-6-mm unwashed coal in any proportion to achieve different product ash levels for different clients.

Table 3 presents the design data for this application. It should be noted that in practice the size range treated in the jig can vary widely. Typical feed size ranges from 5 mm to 80 mm.

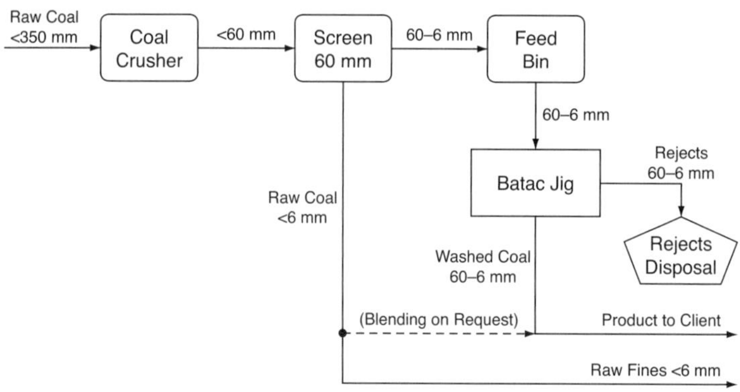

FIGURE 9 Block flowsheet of the Leeuwpan operation

TABLE 3 Design data for the Leeuwpan operation

Data	Design
Feed material	Raw coal
Feed rate (maximum)	389 t/h
Feed ash	>50%
Rejects in feed	50%–60%
Expected cut point	1.8–2.0 t/m^3
Feed size	60–6 mm
Near-gravity material in feed	6%–7%
Feed distribution to the jig	Even in terms of size and mass
Imperfection	0.12

SEPARATION OF COARSE COAL AT LOW CUT POINTS BY DENSE-MEDIUM BATH SEPARATORS (TESKA)

Several dense-medium vessels or baths of proprietary design are available. In light of the limited available space, this discussion considers only the Teska separator because it is in many ways typical of dense-medium baths.

The Teska is a separator for separating raw coal in the size range of 150–10 mm (Figure 10). The separator consists of a slow-moving bucket wheel that operates around a fixed, open-bottom tank, which contains the medium. The bucket wheel is supported on four trunnions and has a roller chain wrapped around and fastened to its outer shell. The entire bucket wheel assembly is rotated by a sprocket mounted on one of the trunnion shafts. The separator and drive are supported on an integral base. The inner fixed portion is fitted with a rubber seal, which is adjusted to a slight clearance from a mating rotation ring on the bucket wheel.

The medium pool is also contained by the bucket wheel, which is fitted with regularly spaced, adjustable orifices to permit a regulated portion of the medium (typically 15%–30% of total flow) to be discharged for return to the circulating medium sump. This discharge produces a controlled down-current in the bath, set to be slightly higher than the settling velocity of the magnetite, which prevents the formation of a high-gravity zone in the dense-medium body. This in turn prevents formation of layers of suspended material, which in some other separators leads to misplaced material. The down-current also accelerates settling of the sinks. This is of special importance in the separation of coal with a large mass of near-gravity material, as is often the case in low-density separations.

The controlled medium downflow described earlier and other design features such as the feed medium distribution manifold and floats paddle ensure that the Teska separator is capable of very sharp separation at very low cut

1. Bucket Wheel for Sinks
2. Perforated Plates
3. Heavy-Media Nozzles
4. Separating Compartment
5. Sinks Discharge Chute
6. Float Discharge Device
7. Drive
8. Support
9. Guide Rollers
10. Drive for Discharge Device
11. Pneumatically Fitted Sealing
12. Feed To Be Washed
13. Dense Medium
14. Sinks
15. Floats

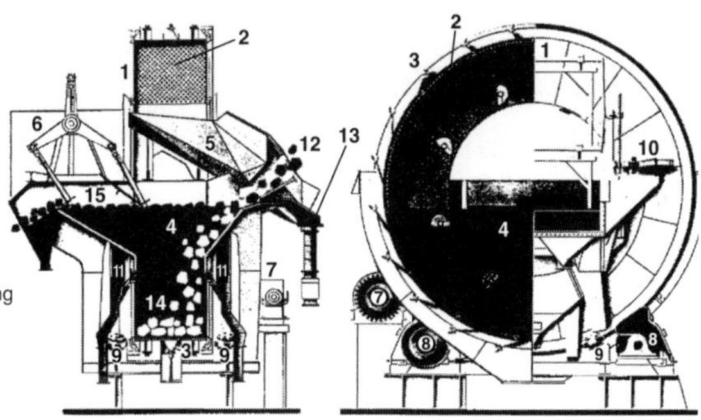

FIGURE 10 Teska dense-medium separator

points. Some points of difference relative to other dense-medium separators are the following:

- The Teska separator is easily accessible. Nothing hinders access, and there is an unhindered view of the separating compartment and the floats discharge.
- The sinks discharge chute can be positioned at the right or left and either concurrent with or opposite to the direction of feeding. This affords maximum layout flexibility.
- A small quantity of medium exists in the bath.
- Minimal maintenance is possible because the drive components do not come into contact with the highly abrasive medium. There are no chains dragging sinks out of the bath.
- Low circulation of dense medium occurs relative to throughput.
- The separator can discharge 100% of feed to the sinks.

Depending on feed size and washability characteristics, the throughput of an individual Teska dense-medium separator ranges up to about 650 t/h. Although the nominal top size limit is 150 mm, the separator can accommodate individual particles up to 250 mm.

The ratio between the volume of circulating dense medium (m^3/h) and raw feed throughput (t/h) typically is in the range of 1.1–1.4. With a coal relative density (RD) at 1.6, this corresponds to a volumetric medium to coal ratio range of 1.8–2.2. This compares to the typical dense-medium cyclone's volumetric medium-to-coal ratio range of 3.5–4 for well-designed plants. This type of process is clearly effective in terms of energy use and wear rates. Typical operating data for Teska dense-medium separators are given in Table 4.

TABLE 4 Operating data for Teska dense-medium separators

Plant	Feed Size, mm	Capacity, t/h	Dense-Medium Circulation, m^3/h	Cut Point, g/cm^3	Ep*
Van Dyks Drift, South Africa	32–6	380	420	1.50	0.013
Fan ge Zhuang, China	150–10	420	500	1.40	0.021
Heinrich Robert, Germany	150–10	480	580	1.40	0.024
Gordonstone, Australia	150–10	200	360	1.40	0.020
Moranbah, Australia	150–10	350	480	1.40	0.025
Dartbrook, Australia	150–10	500	580	1.30	0.025

*Ep = probable error.

SEPARATION OF COARSE COAL BY LARGE-DIAMETER DENSE-MEDIUM CYCLONES

In the last three decades or so, the maximum size of dense-medium cyclones available commercially has roughly doubled, and there are now many 1,500-mm-diameter units in operation. The main impetus for the development of larger cyclone sizes has been the drive for single-train plant modules of increased capacity.

A consequence of going to the larger-diameter units is that the feed, overflow, and underflow openings are also bigger, and therefore there is the potential to pass larger sizes than the traditional 50-mm top size. How big a size can be accommodated depends primarily on the size of the underflow opening (usually the smaller of the two outlets) and the design assumptions used to develop the cyclone geometry, particularly in regard to the probability of multiple top-size particles reporting to the outlet at one time. With reference to the latter, there does not seem to be a universally accepted approach. A common guideline for piping design would be a ratio of maximum size particle to opening of 1:3.

To complicate matters, the design and dimensions of large dense-medium cyclones fall into the following two types:

- Standard-capacity models, generally based on the Dutch State Mines rules and featuring an overflow cap with overflow discharge at right angles to the cyclone axis.
- High-capacity models with elongated barrels, bigger openings, and no overflow cap so that the vortex finder discharges the overflow along the cyclone axis. Figure 11 illustrates the 1,450-mm Multotec unit dubbed the MAX 1450.

Courtesy of Multotec Process Equipment.

FIGURE 11 The Multotec 1,450-mm-diameter dense-medium cyclone

TABLE 5 Coal feed capacity and top size for Multotec dense-medium cyclones

	Multotec Standard-Capacity Cyclones		Multotec High-Capacity Cyclones	
Cyclone Diameter, mm	Maximum Particle Size, mm	Coal Feed, t/h	Maximum Particle Size, mm	Coal Feed, t/h
510	34	54	51	99
610	41	81	61	145
660	44	97	66	175
710	47	114	71	207
800	53	149	80	270
900	60	196	94	355
1,000	67	249	100	454
1,150	77	351	115	638
1,300	87	468	130	854
1,450	97	608	145	1,108

Courtesy of Multotec Process Equipment.

Large units are now available from several cyclone suppliers. It is understood that the feed top size to the larger cyclones is set at about 100 mm or more.

By way of illustration, Table 5 presents relevant coal feed capacity and maximum feed size for the two types of cyclones available from Multotec in South Africa. These data are based on the following:

- Equivalent feed head of nine times the nominal cyclone diameter
- Volumetric medium-to-coal ratio of 3.5:1
- Feed coal solids of 1.6 RD

Figure 12 presents the Multotec recommended feed size ranges by dense-medium cyclone diameter. For the MAX 1450 unit referred to previously, the recommended maximum feed top size is 145 mm.

A search of the literature has failed to locate any dense-medium cyclone applications with a feed top size greater than 50 mm. One can only speculate as to the reason. On one hand, there are practical considerations such as the formidable pumping task. For example, a Multotec high-capacity 1,450-mm-diameter dense-medium cyclone would need a pump capable of delivering 3,059 m^3/h of pulp at a discharge head sufficient to achieve 20 m of head at the cyclone feed inlet and with an impeller capable of passing 100-mm particles. A comparable circulating flow for a dense-medium bath would be about half as large, at far less feed head, and would require handling of only medium solids, so that the more efficient closed type of impeller could be used.

On the other hand, there is the question of separation performance. A recent investigation by Swanson and Atkinson (2007) into the performance of large-diameter (1,150-mm and 1,300-mm) dense-medium cyclones on nominal minus-50-mm feed in three Australian plants points to a significant

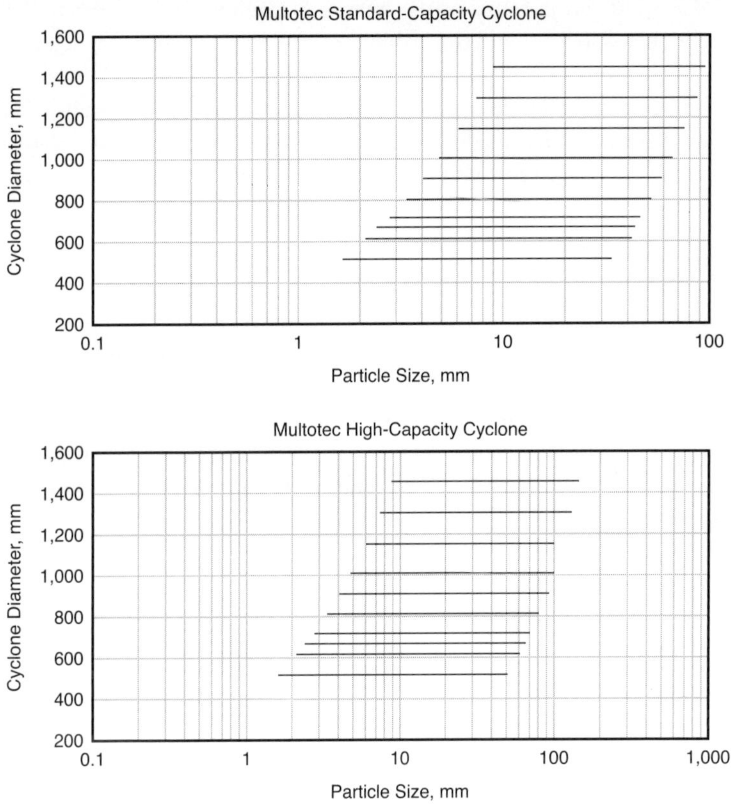

Courtesy of Multotec Process Equipment.

FIGURE 12 Multotec recommended feed size by dense-medium cyclone diameter

performance deterioration in terms of imperfection value (defined as Ep/RD50) at sizes below about 10 mm. It was also shown that medium-to-coal ratios above 3:1 are needed. A pattern of low-RD coal losses to the underflow was also noted.

These findings followed work by Multotec that identified a "breakaway" size, d_b, where performance in terms of imperfection value deteriorated. Multotec recommend that no more than 10% of feed mass should be below d_b, depending on the difficulty of coal separation in this size fraction. For the MAX 1450 unit mentioned earlier, the value of d_b is 9.2 mm, so the recommended feed size range would be 145–9.2 mm in nominal terms.

In summary, the largest dense-medium cyclones available commercially can treat a feed size range comparable to a dense-medium bath. However, they have disadvantages in terms of the pumping requirements, accelerated wear due to higher velocities, and the inability to pass rogue oversize particles.

It is therefore reasonable to think that where washability characteristics are appropriate, plant designers will continue to opt for separate dense-medium bath and dense-medium cyclone circuits with the flexibility of setting two cut points so as to best tailor the separation task over the full size range.

CONCLUSIONS

A steady feed rate is essential to any separation operation. Romjig, Batac jig, dense-medium baths, and dense-medium cyclones are no exception.

Correct sizing of the feed is important. Raw coal should be prescreened; screening at required sizes should present no problems. Excessive amounts of oversize and undersize impede the separation process.

The elimination of stone at large sizes (that is, from ROM or raw coal, before crushing) substantially reduces the need and cost to crush stone. It also reduces the amount of tailings and fines exiting the washery, thus reducing the moisture content of the final product.

Placed immediately ahead of a washery, de-stoning represents a capacity increase at a low unit cost and with minimum disruption to existing operations. Alternatively, by de-stoning in a near-pit location—made feasible because of the limited water and operator attention needed—it is possible to decrease the transport cost of ROM coal to the washery and reject to disposal. This is particularly relevant to older open-cut mines with long truck cycle times.

Trying to save money in the wrong places in plant design may prove to be very costly in the end. Coarse coal separation plants save money. Easy separation at large sizes, less maintenance, reduced moisture in the washed coal, and the high reduction of fines generation in the total plant process are potential arguments for the cost-effective coal washing plant of the future.

REFERENCES

Kottmann, J., Yannoulis, G.F., and Ziaja, D. 2006. De-stoning—The way to improve coal mine economics in modern coal preparation. In *Proceedings of the XV International Coal Preparation Congress and Exhibition*, Beijing. p. 442.

Sanders, G.J., and Ziaja, D. 2003. The role and performance of Romjig in modern coal preparation practice. In *Proceedings of CoalPrep '03*, Lexington, KY, April 29–May 1. p. 205.

Sanders, G.J., Gnanaiah, E.U., and Ziaja. D. 2000. Application of the Humboldt de-stoning process in Australia. In *Proceedings of the Eighth Australian Coal Preparation Conference*, Port Stephens, NSW, November. Paper A2.

Sanders, G.J., Ziaja, D., and Kottmann, J. 2002. Cost-efficient beneficiation of coal by Romjigs and Batac jigs. In *Proceedings of the XIV International Coal Preparation Congress*, Johannesburg, Republic of South Africa, March. pp. 395–400.

Swanson, A., and Atkinson, B. 2007. Further operational data to optimise the application of large diameter dense medium cyclones. In *Proceedings of the 11th Australian Coal Preparation Conference*, Twin Waters, Queensland, Australia, March. Paper G2.

Wesp, M., Thelen, H., Müller, H., and Wüster, G. 1989. Operational experience on a moving bed jig built by KHD Humboldt Wedag AG for pre-destoning at the Emil Mayrish colliery. *Aufber.-Techn.* 30(12):746–752.

Ziaja, D., and North, A. 1998. Cost efficient beneficiation of coking and steaming coal. *Aufber.-Techn.* 39(11):553–560.

Dense-Medium Cyclones

G.J. de Korte and J. Engelbrecht

ABSTRACT

The dense-medium cyclone, first developed by the Dutch State Mines (DSM) in the Netherlands in the 1940s, has firmly established itself as the processing unit of choice in many minerals industries. Application of the dense-medium cyclone ranges from the recovery of diamonds, iron ore, and chromite at high relative densities (RDs) through the separation of plastics at RDs below 1.

Despite its simple design, the dense-medium cyclone is not fully understood, and work aimed at improving knowledge about cyclone behavior continues. The main focus of further development is intended for making cyclones more efficient and increasing throughput capacity. The performance of dense-medium cyclones is controlled by many parameters, and the influence of some of these factors is discussed in this chapter.

CYCLONE DIMENSIONS

Recent times have seen cyclone diameters increasing to the point where 1,500-mm-diameter cyclones are being used in many coal processing applications, particularly in Australia. The larger diameter also implies larger feed, overflow, and underflow openings and thus allows larger particles to be fed to cyclones. DSM standard dimensions and current manufacturing trends for dense-medium cyclones are defined in Table 1.

Most cyclone manufacturers still adhere to the DSM recommendations, and Multotec "standard" cyclones are manufactured to these dimensions. However, there is a demand for cyclones having higher capacity, the ability to process larger feed particles, and the ability to handle higher amounts of sink material. The latter requirement applies because modern mining methods are less selective and include more roof and floor shale in the coal mined and because more low-grade reserves are being mined. In India, for example, raw coal contains fairly large proportions of high-density material. As

TABLE 1 Standard cyclone dimensions

Parameter	DSM Recommendations	Current Manufacturing Trends
Cyclone diameter		Up to 1,500 mm
Inlet size	0.2 × cyclone diameter	0.2, 0.25, or 0.3 × cyclone diameter
Vortex finder diameter	0.43 × cyclone diameter	0.43 or 0.50 × cyclone diameter
Barrel length	0.5 × cyclone diameter	0.5 to 2.5 × cyclone diameter
Spigot diameter	0.3 × cyclone diameter	0.3 to 0.4 × cyclone diameter

TABLE 2 Dimensions and capacities of Multotec cyclones

	Standard-Capacity Cyclones		High-Capacity Cyclones	
Cyclone Diameter, mm	Maximum Particle Size, mm	Coal Feed, t/h	Maximum Particle Size, mm	Coal Feed, t/h
510	34	54	51	99
610	41	81	61	145
660	44	97	66	175
710	47	114	71	207
800	53	149	80	270
900	60	196	94	355
1,000	67	249	100	454
1,150	77	351	115	638
1,300	87	468	130	854
1,450	97	608	145	1,108

a result of these requirements, cyclone dimensions are changing to reflect the current trends shown in Table 1.

Table 2 presents relevant coal feed capacity and maximum feed size for the two types of cyclones available from Multotec (South Africa). The values given in Table 2 are based on the following:

- Equivalent feed head of nine times the nominal cyclone diameter
- Volumetric medium to coal ratio of 3.5:1
- Feed coal solids relative density (RD) of 1.6

A 1,450-mm-diameter Multotec cyclone is shown in Figure 1.

Although the larger cyclones can handle large feed particles up to 140 mm, these particles can cause hang-up problems. Cyclones normally operate with a typical feed top size of 50 mm. This is probably because of the difficulty in pumping large particles to the cyclone.

BREAKAWAY SIZE AND CUT-DENSITY SHIFT

The breakaway size is defined as the particle size below which the recovery efficiency for the smaller particles starts to decrease significantly. Bosman (1994) provides an approximate breakaway size versus cyclone diameter, as

FIGURE 1 Multotec 1,450-mm-diameter dense-medium cyclone

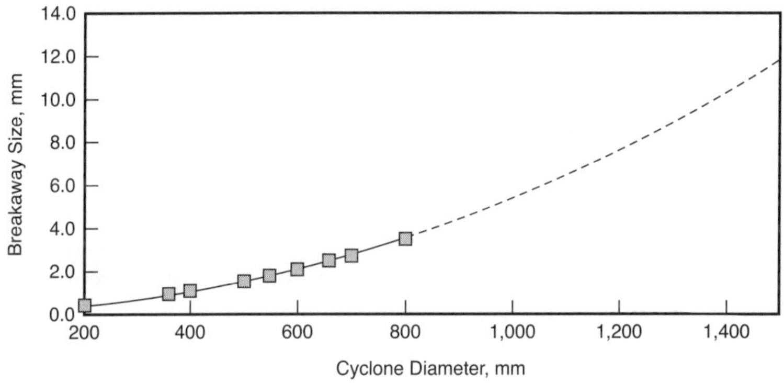

Source: Bosman 1994.

FIGURE 2 Breakaway size versus cyclone diameter

shown in Figure 2. Bosman includes cyclones only up to 800 mm in diameter, and the graph has been extended to represent cyclones of 1,500-mm diameter in the figure.

One problem with the breakaway size definition is it indicates only that the efficiency decreases significantly, without quantifying this decrease. The probable error (EPM [écart probable moyen]) is a function not only of cyclone diameter but also of feed pressure, medium density, medium viscosity, particle shape and density, top size of particles in the feed, and so forth. Actual data from a number of South African operations

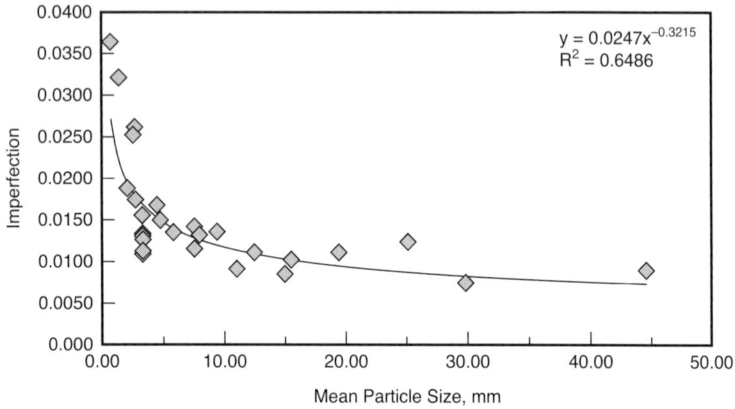

FIGURE 3 Particle size versus imperfection for South African cyclones

(de Korte 2007a), using mostly 610-mm-diameter cyclones, are summarized in Figure 3, which indicates that the imperfection does deteriorate significantly as the size of feed particles gets smaller.

Defining the breakaway size has led some people to believe that dense-medium cyclones are not capable of beneficiating coal that is smaller than the breakaway size and that the minus-3-mm size fraction, for example, should be screened out of the cyclone feed and processed with water-only units such as spirals or teeter-bed separators (TBSs). It has also contributed to the industry being reluctant to adopt large-diameter cyclones. In reality, the efficiency does drop off as particles become smaller, but the efficiency obtained in dense-medium cyclones is still much better than that obtainable from spirals or TBSs. Table 3 shows the comparative efficiency data (de Korte and Bosman 2006) for dense-medium cyclones and water-only units.

Simultaneously with the change in efficiency as a function of particle size for a specific cyclone diameter, there is a concurrent shift in the cut density (SG_{50}). Figure 4 shows a typical result for a 610-mm-diameter cyclone where both the normalized EPM and the relative cut density are given as a function of particle size.

It is important to note that the shift in cut density for water-based separators such as jigs, spirals, and hindered settling devices is more pronounced for jigs in comparison with dense-medium cyclones, as shown in Figure 5.

TABLE 3 Dense-medium cyclones versus water-based units

	Medium Based						Water Based		
Parameter	800-mm Cyclone	660-mm Cyclone	500-mm Cyclone	500-mm Cyclone	200-mm Cyclone	Two-Stage Cyclone	Two-Stage Spirals	Reflux	TBS
Size, mm	3 × 0.5	2 × 0.5	3 × 1	1 × 6	3.35 × 0.212	1 × 0.1	1 × 0.1	3 × 0.5	3 × 0.5
EPM	0.043	0.079	0.021	0.031	0.026	0.054	0.150	0.101	0.104
Cut density (SG_{50})	1.563	1.750	1.312	1.380	1.350	1.743	1.800	1.630	1.672
EPM/SG_{50}	0.028	0.045	0.016	0.022	0.019	0.031	0.083	0.062	0.062

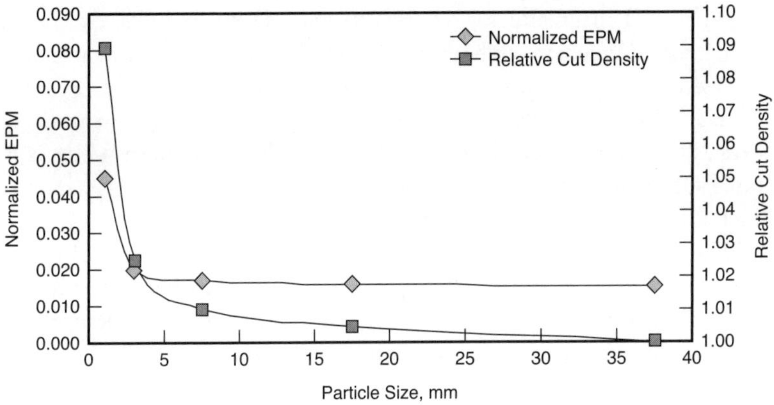

FIGURE 4 Particle size versus normalized EPM and relative cut density

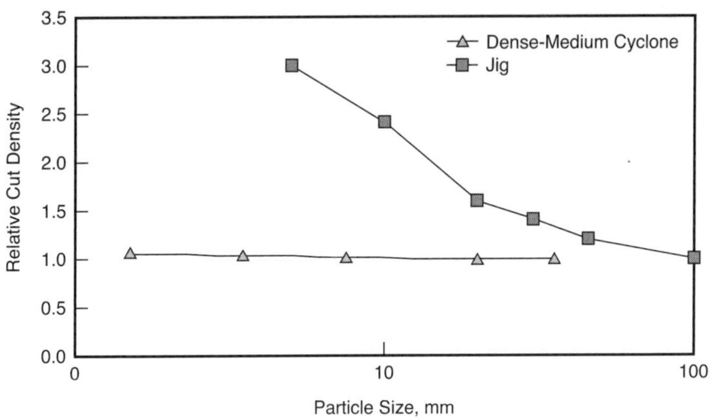

FIGURE 5 Particle size versus relative cut density for dense-medium cyclones and jigs

TABLE 4 DSM-specified magnetite size grading

Density of Circulating Medium	Magnetite Grade	Grain Size of Magnetite
1.25–1.40	Grade F	95% <40 microns
1.40–1.90	Grade E	95% <50 microns

MEDIUM

The size-consist and viscosity of the medium used in the dense-medium cyclone are of critical importance in the unit's operation. The influence of the size-consist of the medium on the operation is fairly well understood, and guidelines have been specified by DSM (Anonymous, n.d.). They recommend fine (Grade F) magnetite for low-density separations and slightly coarser Grade E for intermediate- to high-density separations, as shown in Table 4.

DSM also indicated that the cyclone performance deteriorates when the magnetite used becomes coarser. The results of some recent tests (de Korte 2007b) conducted at South Witbank Colliery in South Africa confirmed DSM findings. The influence of the percentage minus-40-micron material in the medium on imperfection is shown in Figure 6.

The medium in the dense-medium circuit does not necessarily have the same size-consist as that of the medium purchased because the finer size fractions of the medium are not recovered as effectively by the magnetic separators as the coarser size fractions. The result is that the in-circuit medium gradually becomes coarser.

The correct medium specification therefore depends on the viscosity of the medium and the centrifugal force in the cyclone. The variables that determine these properties are

- Size distribution of the magnetite
- Feed density
- Percentage of slimes in the medium
- Cyclone dimensions
- Operating pressure

Contamination of the medium with nonmagnetic material, especially by ultrafine slimes, is known to increase the viscosity of the medium and may decrease separation efficiency. For this reason, it is important to keep the medium clean by constantly bleeding some of the medium to the magnetic separators to remove the nonmagnetic materials. Bleeding is also needed to control the medium density. Because an increase in medium density is associated with a reduction in the volume of medium in the correct density tank, many plant operators believe that medium is lost by bleeding to the magnetic separators. Operators therefore tend to keep this at an absolute minimum, which often results in contaminated medium.

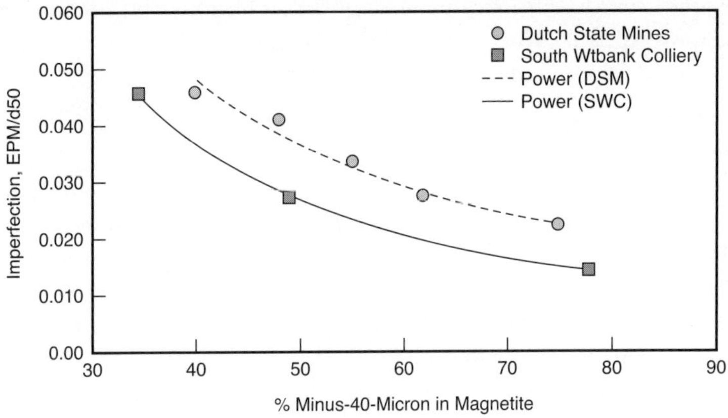

FIGURE 6 Influence of magnetite size-consist on imperfection

A new patented flowsheet by Multotec uses a high-intensity magnetic drum separator that gives maximum recovery of the medium at high densities, which minimizes losses of medium and removes contamination. For example, densities of 85% solids (RD = 3.1) were achieved on a fine magnetite with 99.9% recovery.

The effect of demagnetizing coils on the operation of dense-medium cyclones is not well known. In general the belief is that demagnetizing coils are not necessary in cyclone circuits because the high centrifugal forces in the cyclone counteract any tendency toward magnetic flocculation of the medium. Most cyclone circuits therefore are not equipped with demagnetizing coils. Napier-Munn and Scott (1990) found that the viscosity of the medium in diamond recovery circuits was lowered by passing of the ferrosilicon recovered by the magnetic separators through a demagnetizing coil.

Test work at a coal mine with low separation densities showed that demagnetizing the media lowered the differential between the feed and the cut density. This implies that without demagnetizing, some of the magnetic flocs survive the shear forces in the cyclone.

FEED DISTRIBUTION

Where more than one cyclone in parallel is needed, it is necessary to divide the feed equally between two or more cyclones. Almost all the designs of feed distributors installed in processing plants fail in this regard when they are not properly designed, taking the whole pipe system into consideration. The consequence is that cyclones are unevenly fed. This results in a reduction of separation efficiency and is one of the main motivating factors for replacing many small cyclones with a single large-diameter unit.

CONTROL

Where several cyclones are operating as a single module, the cyclones often cut at slightly different densities. This may be because the cyclone dimensions are slightly different, especially the spigot diameters, because of uneven wear or selective renewal of some spigots, as a result of uneven feed distribution or poor density control when the cyclones are fed with separate medium circuits. In all cases, this results in a loss of efficiency.

At a coal preparation plant with eight 500-mm cyclones running in parallel, the misplaced material increased from 6% to 13% only because different spigot sizes were used, which meant that the cyclones were cutting at different densities.

When feed with high amounts of near-density material is being processed, the control of the medium density is critical. Very small fluctuations in the density can cause dramatic changes in the product yield and the amount of misplaced material. This is illustrated in the well-known graph of Horsfall (1987), where the effect on organic efficiency as a function of the EPM for separation of coals with different percentages of near-density material is shown (Figure 7).

FEED PRESSURE

As recommended by DSM, most dense-medium cyclones in the coal industry are operated at a feed pressure of 9 diameters. The feed pressure is a very important parameter in that it has a significant influence on cyclone performance. Recent work done by Steyn (2006) shows that the efficiency of a 900-mm-diameter cyclone was appreciably improved by an increase in feed pressure. The effect of an increase in feed pressure on the EPM of the cyclone is shown in Figure 8.

It seems that the beneficiation of the coarser feed particles is preferentially improved by a feed pressure increase. Figure 9 shows that the imperfection in the fine coal does not change appreciably with a pressure increase, whereas the imperfection is reduced for the larger coal sizes.

Increasing the feed pressure increases the throughput capacity of the cyclones but also increases wear in the cyclone and hence decreases the useful lifetime of the cyclone. Therefore, feed pressure must be considered carefully in operation design.

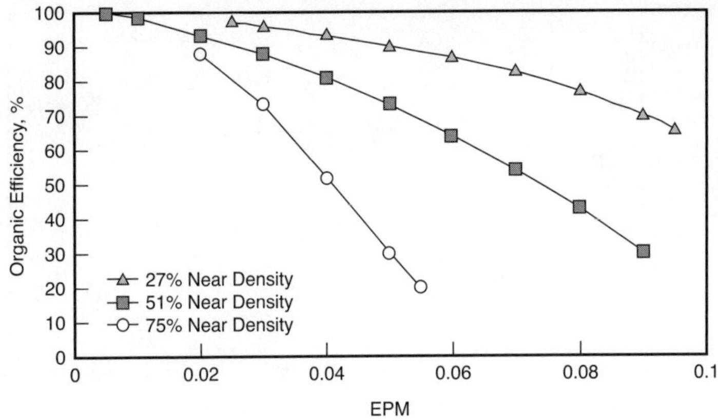

FIGURE 7 Organic efficiency versus EPM for various percentages of near-density material

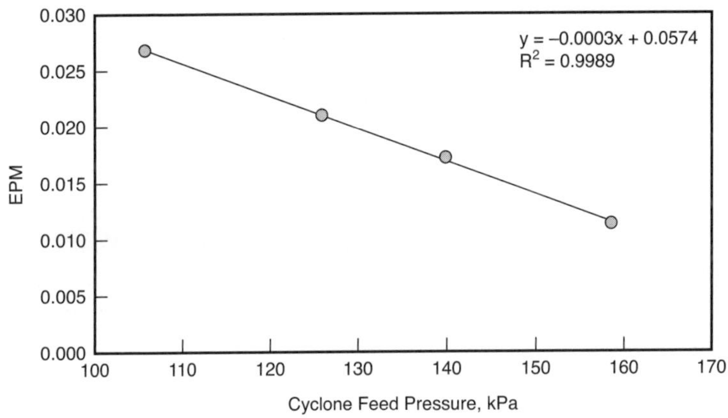

FIGURE 8 Cyclone feed pressure versus EPM for 900-mm cyclone

APPLICATIONS

Cyclones were traditionally used to process material between approximately 20 mm and 0.5 mm. The advent of large-diameter cyclones now allows coarser material to be processed. However, the problems associated with pumping large particles still limit the upper size of material processed in dense-medium

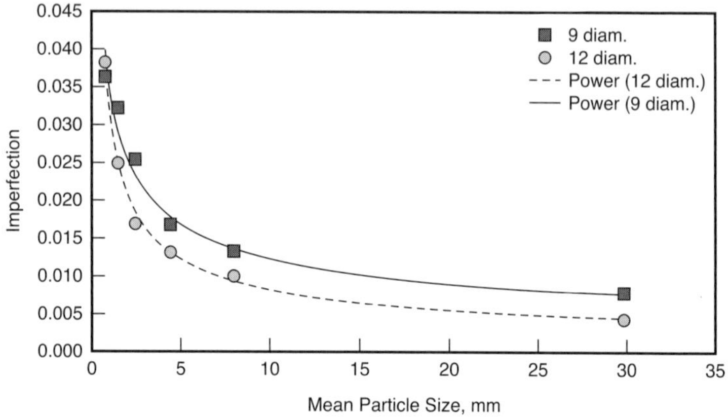

FIGURE 9 Imperfection versus particle size at various cyclone feed pressures for a 900-mm cyclone

cyclones to approximately 50 mm. Where sized products are needed, it is still customary to install a dense-medium bath and to process only the small material via dense-medium cyclones.

On the lower end of the size scale, 0.5 mm was always considered the practical limit for dense-medium cyclones. However, there have been significant efforts to process coal finer than 0.5 mm with dense-medium cyclones. The plants built at Winterslag and Tertre in Belgium, Homer City and Marrowbone in the United States, Greenside in South Africa, and Curragh in Australia serve as proof. Unfortunately, the results obtained from these plants were not always as good as expected. Recent advances in magnetic separators have prompted renewed interest in using the dense-medium cyclone for fine coal processing. The efficiency of separation obtainable with dense-medium cyclones is much better than that of water-only units such as spirals and TBSs. For this reason, a new fine coal dense-medium plant is under construction in South Africa. The change that enabled the successful use of larger-diameter cyclones and coarser media for fine particle dense-medium separation was multistage treatment of fine coal, which resulted in an overall efficient process (de Korte 2002).

By virtue of their high separation efficiency, dense-medium cyclones are the method of choice for processing difficult-to-process raw coals. They are also applied in cases where the price of the product dictates that the highest possible yield be obtained. The growing number of dense-medium cyclone plants being used in countries such as India and the extensive use of cyclones in the United States with easy coals are proof of this.

REFERENCES

Anonymous. n.d. The heavy medium cyclone washery for minerals and coal. In *Dutch State Mines (DSM) Handbook*. The Netherlands: Dutch State Mines.

Bosman, J. 1994. Dense medium cyclones: Sizing and selection. Presented at the 5th Symposium on Dense-Medium Separation, Bongani Sun Lodge (Samancor), South Africa, September.

de Korte, G.J. 2002. Dense-medium beneficiation of fine coal revisited. Presented at the XIV International Coal Preparation Congress, Johannesburg, South Africa, March 11–15.

———. 2007a. *Processing Fine Coal in Large Diameter Cyclones*. CSIR Report No. 2007-0062. Pretoria, South Africa: Council for Scientific and Industrial Research.

———. 2007b. *Tests Conducted by Martin and Robson to Evaluate the Potential of Coarse Magnetite for Use in Coal Processing*. CSIR Report No. 2007-0031. Pretoria, South Africa: Council for Scientific and Industrial Research.

de Korte, G.J., and Bosman, J. 2006. Optimal processing route for the 3 × 0.5 mm size fraction? Presented at the DMS and Gravity Concentration Operations and Technology in South Africa Conference, Pretoria, South Africa, July 18–20.

Horsfall, D.W. 1987. The lows and highs of dense medium separation. Presented at the Dense Medium Operators' Conference, AusIMM, Southern Queensland Branch, July.

Napier-Munn, T.J., and Scott, I.A. 1990. The effect of demagnetisation and ore contamination on the viscosity of the medium in a dense medium cyclone plant. *Miner. Eng*. 3(6):607–613.

Steyn, J. 2006. The Effect of Inlet Pressure on Dense Medium Cyclone Performance, Greenside Colliery, South Africa. Unpublished report.

Design of High-Efficiency Spiral Circuits for Preparation Plants

Gerald H. Luttrell, Rick Q. Honaker, Peter J. Bethell, and Fred L. Stanley

ABSTRACT

Spiral separators, which were first introduced to the coal industry in the early 1980s, have become one of the most popular choices for cleaning fine coal in the 1 × 0.15 mm size range. The popularity of this simple process can be attributed to several factors, including ease of operation (no moving parts), low capital and operating costs, and inherently low coal losses. Despite these advantages, field studies suggest that coal spirals have performed poorly in some early installations as a result of poor feed distribution, excessive flow rates, sanding and beaching of solids, and improper splitter settings. Furthermore, the use of single-stage spirals in many older plants often results in the misplacement of high-ash rock into the clean coal product. In response to this inherent shortcoming, spiral manufacturers now offer compound spirals that incorporate an additional stage of cleaning or scavenging along the same unit. The multistage design, when combined with proper operating conditions, provides exceptionally good levels of coal cleaning performance. This chapter reviews some of the important guidelines that must be followed to ensure that spiral circuits are properly designed, operated, and maintained.

INTRODUCTION

The invention of spiral separators is widely credited to Ira Boyd Humphreys, an award-winning inventor who developed the Humphreys spiral during World War II for concentrating heavy mineral sands (Colorado Historical Society 2001). Most early spiral designs were constructed from cast iron or cast concrete. These units yielded a low throughput per unit equipment weight and therefore were impractical for separating most low-value materials in the mining industry. This situation changed in the 1960s with the commercial development of lightweight fiberglass spirals that could be arranged in twin- and triple-start assemblies. However, it was not until the early 1980s that spirals were designed specifically for the coal industry (Weldon 1997). Compared with mineral spirals, this new generation of coal spirals was manufactured with a lower pitch to reduce velocity and increase residence time to enhance the separation of low-density coal particles. In addition, coal spirals were manufactured with a larger diameter so that larger tonnages of solids could be processed by each unit (Osborne 1988).

OPERATING PRINCIPLE

A typical spiral consists of an elongated helical trough with an elliptical cross-section (Figure 1). Two- or three-spiral units are commonly intertwined along a single support "stick" to improve throughput and save floor space. These clusters are generally incorporated into banks that commonly include 18 or more individual spiral units that are fed from overhead distributors. Slurry from each leg of the distributor is introduced to the start of each spiral and

FIGURE 1 Bank of triple-start coal spirals

allowed to flow by gravity downward along the corkscrew-shaped path of the spiral. The resultant flow pattern subjects particles to a variety of forces that cause lighter coal particles to move to the outer wall and denser particles to move toward the center. Once the particles become segregated across the spiral profile, adjustable splitters or cutters are used to divert different sections of the flowing streams to the clean coal, middlings, or refuse products.

The sorting phenomenon that occurs as slurry flows along the length of the spiral has been the source of much confusion in the literature. The flow of slurry passing down the length of the spiral run is easy to visualize. The high-velocity region near the outer radius of the trough accounts for most of the tangential fluid flow. Ideally, most of the clean coal is pushed into this region by the radial flow patterns across the trough profile. The tangential flow thins near the center of the spiral and is slowed by the skin and drag forces. Particles of rock, which are denser, typically are forced into this region by differential settling forces. On the other hand, conceptual drawings of spiral trough cross-sections suggest that the radial liquid flow sweeps upward toward the outer radius of the spiral and then returns underneath, along the surface of the trough (Figure 2a). This radial flow pattern is assumed to carry lighter particles of coal outward into the high-velocity flow zone, where they can be collected separately from the refuse that is pushed toward the center of the spiral.

This description fails to recognize that two counterrotating flows actually exist across the profile of a spiral trough (Figure 2b). The two rotating flow patterns converge at a single point where the surface flow folds in on itself. Objects lighter than water (e.g., wood and air bubbles) move to this point on the surface of the liquid and remain there for the entire ride down the spiral (Figure 3). The clockwise flow in the lower rotation zone carries lighter particles of coal back to the outside wall of the spiral, where they are recovered, whereas heavier particles settle and are carried to the inside of the spiral, where they are rejected. This action allows spirals to provide a good refuse product that is largely free of coal. In contrast, the counterclockwise flow in the upper rotating section stratifies the particle bed along the outside wall so that particles of different densities can be further segregated. Unfortunately, particles of rock that enter the upper zone tend to settle against the wall and are held there by the upward flow. As a result, it is very difficult for these particles to pass across the flow separation line and be rejected. These high-ash particles eventually report to the low-density stream and contaminate the coal product. The twin-rotating flow pattern provides an explanation as to why spirals can suffer from a high specific gravity (SG) cut point and a tendency to misplace rock into the clean coal product (M. Mankosa, personal communication).

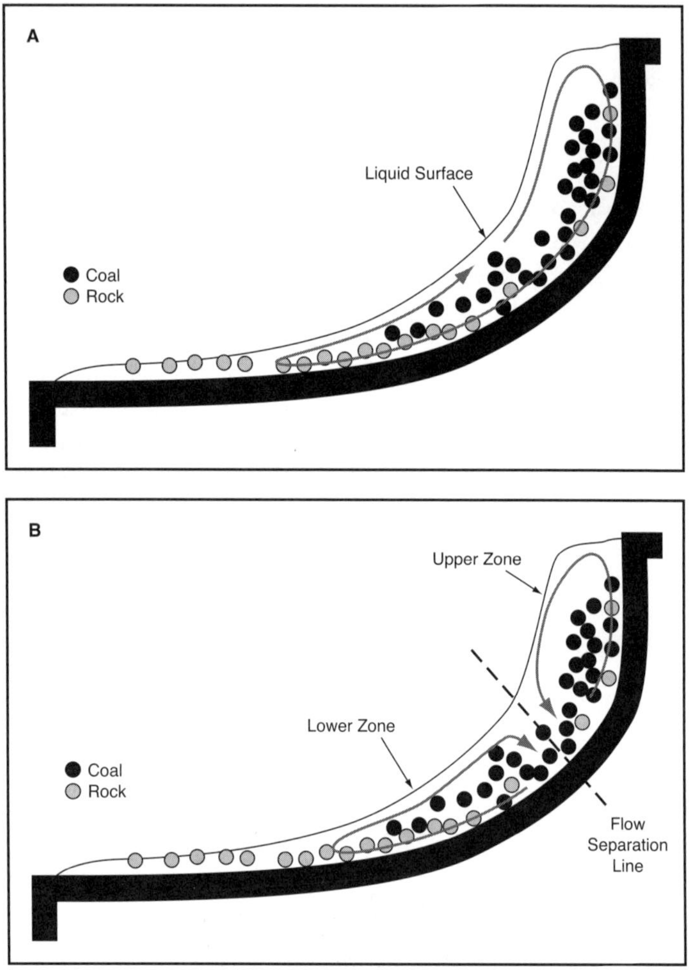

FIGURE 2 Schematics showing (a) the incorrect flow pattern historically assumed for spirals and (b) the counterrotation flow pattern that actually exists

SPIRAL COMPARISON

There are several commercial manufacturers of coal spirals, including Mineral Deposits, Multotec, Carpco/Outokumpu, Krebs/SWMS, Minpro, and Linatex. Some of the manufacturers claim to incorporate proprietary features into their spiral designs that are intended to improve separation performance. Although debate about the benefits of these features is likely to continue, many studies suggest that there is actually little difference in the various designs. For example, Figure 4 shows test data obtained from a side-by-side

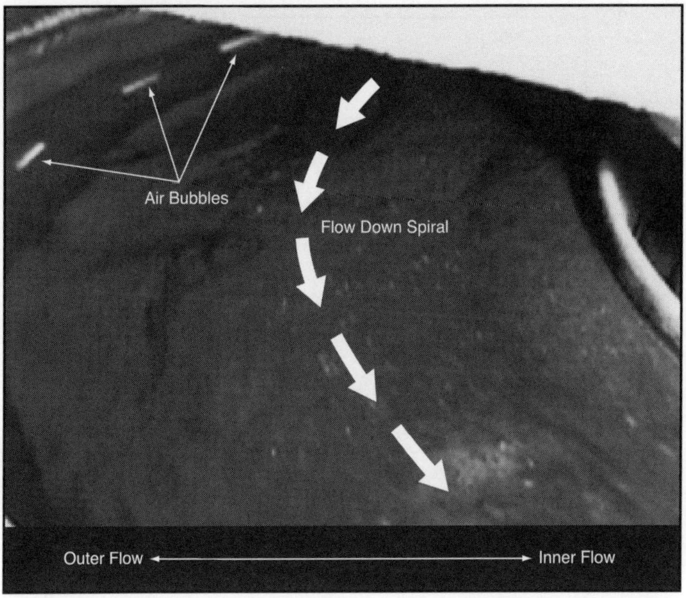

FIGURE 3 Air bubbles riding along the flow separation line down the spiral length

comparison of four single-stage spirals (Honaker and Wang 1991). All the experimental runs were performed using a closed-loop test circuit to ensure that each spiral received the same feed material. During each test series, the volumetric flow rate of feed slurry was varied over three levels to generate a recovery–grade curve for each spiral. The feed solids content was held constant at 30% solids by weight. The data show that there is very little difference in the separation performance offered by different single-stage spiral designs. However, routine in-plant sampling programs do indicate that spiral designs from some manufactures are somewhat less prone to sanding than others. Therefore, these designs may be preferred over the long run in an actual plant environment to ensure that optimal performance is maintained in the absence of operator attention.

TWO-STAGE CIRCUITS

Although only small differences exist in the performance of different brands of spirals, studies have shown that the configuration of spirals in multistage circuits does have a large impact on separation efficiency (Bethell et al. 1991; Bethell 2002; Bethell and Arnold 2003). The middlings product from a spiral typically contains composite (poorly liberated) particles of intermediate SG

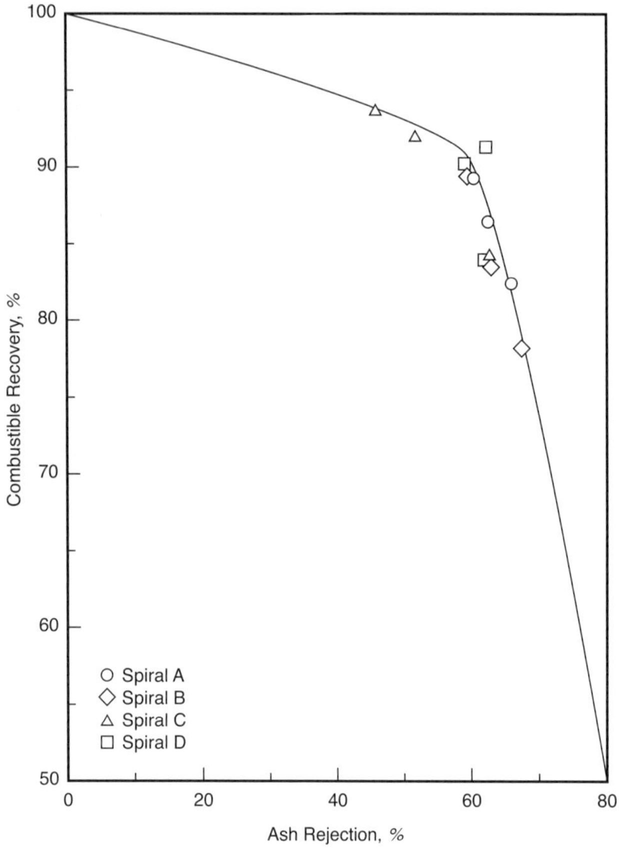

FIGURE 4 Performance comparison for four single-stage commercial spirals

and misplaced particles of more pure rock and coal. In a single-stage circuit, operators are forced to discard the middlings (and sacrifice coal yield) or retain the middlings (and accept a lower clean coal quality). To overcome this problem, two-stage circuits have been introduced in which clean coal or middlings are retreated using a second stage of spirals. A common configuration is the two-stage scavenger circuit shown in Figure 5a. This arrangement uses a downstream bank of secondary spirals to rewash middlings produced by the primary spirals. Operators often prefer this arrangement because very few additional spirals are needed to handle the small flow of primary middlings. This circuit does improve separation efficiency, but the additional coal product scavenged by the secondary spirals further raises the SG cut point for the overall circuit. The cut point may become unacceptably high, resulting in a

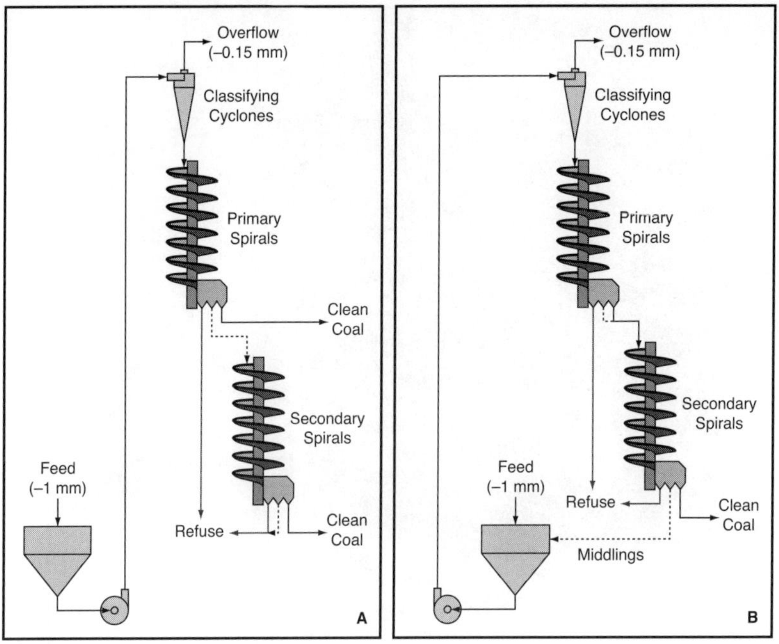

FIGURE 5 Simplified flowsheets for two-stage (a) traditional scavenger circuit and (b) modern cleaner circuit

clean coal product that does not meet quality specifications. This spiral configuration is also incapable of eliminating rock that may be misplaced into the clean coal product during the primary stage of separation. Fortunately, both of these problems can be minimized through the use of a cleaner circuit, shown in Figure 5b. In this configuration, both the clean coal and middlings products are retreated by the secondary spirals. This configuration reduces the SG cut point and rejects a greater proportion of misplaced rock. The only downside to this layout is that substantially more secondary spirals are needed because both the middlings and entire clean coal product must be rewashed. The middlings stream from the secondary spirals typically is recycled to further improve the separation efficiency. The potential improvements in separation efficiency that may be obtained by recycling middlings has been discussed elsewhere (Luttrell et al. 2007). Data from full-scale plant trials showed that the cleaner configuration improved clean coal yield by 6.6 percentage points over a single-stage spiral and by 3.4 percentage points over a two-stage scavenger configuration (Luttrell et al. 1998).

The potential improvement in separation performance afforded using cleaner circuits has led several manufacturers to offer compound units that

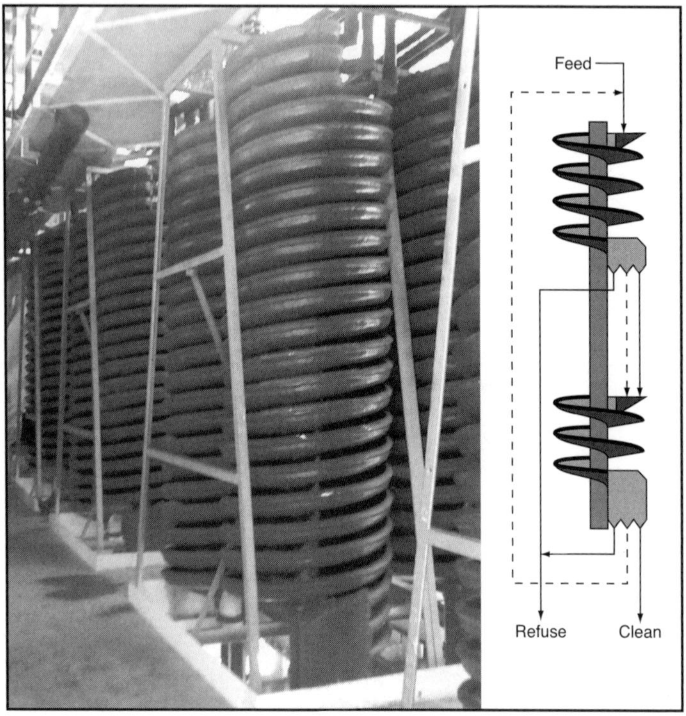

FIGURE 6 Commercial installation of modern compound cleaner spirals (Multotec SX7)

incorporate two stages of spirals in a single low-profile assembly. Typically, these units consist of four turns of primary spirals, followed immediately by three turns of secondary spirals (Figure 6). The refuse from both the primary and secondary spirals is rejected, and clean product is taken only from the secondary spirals. The clean product and middlings from the primary turns are remixed in a pulping box before being fed to the secondary turns. Ideally, the secondary middlings product is recycled back to the primary spiral feed. Although the compound configuration is only marginally more costly than a single-stage circuit, the two-stage process offers large improvements in separation performance. Studies by Bethell and Arnold (2003) showed that, depending on the range of operating variables selected, SG_{50} values of 1.68–1.85 SG and probable error (Ep) values of 0.16–0.18 could be achieved using compound cleaner spirals. As shown in Table 1, the compound cleaner spirals increased the recovery of float 1.6 SG fraction from 96.5% to 97.3% and the recovery of 1.6 × 1.8 SG feed fraction from 25.4% to 54.4% when compared

TABLE 1 Comparison of two-stage scavenger and cleaner spiral circuits

	Two-Stage Scavenger, %			Two-Stage Cleaner, %		
SG Class	Clean	Refuse	Partition	Clean	Refuse	Partition
Float 1.6	96.53	3.87	97.99	97.30	1.40	100.00
1.6 × 1.8	1.01	2.80	25.43	1.65	0.96	54.41
Sink 1.8	2.46	93.33	5.86	1.03	97.63	1.36

with traditional middlings scavenger spirals. The compound cleaner spirals also simultaneously reduced the recovery of the 1.8 SG sink fraction in the clean coal from 5.9% to 1.4%. The superior separation efficiency offered by the cleaner configuration is rapidly making this approach the standard configuration for plant retrofits and new installations.

OPERATING AND DESIGN PARAMETERS

Spirals are capable of maintaining good recoveries of clean coal when correctly configured and properly operated. Unfortunately, spiral circuits are not often run under optimal conditions because of problems such as poor feed distribution, sanding or beaching, improper splitter settings, and incorrect liquid and solid flow rates. Therefore, even well-designed spiral circuits must be properly operated to ensure optimal performance.

Dry Feed Rate

Some spiral manufacturers and plant designers suggest that spirals are capable of achieving capacities of 4.1 t/h (4.5 tph) or more of dry feed solids per start. However, the SG cut point increases rapidly as the feed tonnage to the spiral increases (Figure 7a). Consequently, it becomes difficult to maintain a low SG cut point when operating at high loadings exceeding about 2.3 t/h (2.5 tph) per start. The efficiency of a spiral also generally diminishes as the tonnage rate increases. The Ep can reach unacceptably high values when operating at feed rates above about 2.3 t/h (2.5 tph) per start (Figure 7b). Values lower than 2.3 t/h (2.5 tph) may be justified when cleaner products (lower SG cut points) are needed. Similar trends in cut point and Ep have been reported in the literature (Li et al. 1993; Subsinghe and Kelly 1991, 1992; Holland-Batt 1994). High SG cut points and poor Ep values are common in industrial plants that have installed too few spirals because of constraints associated with limited capital or have been pushed beyond the rated design capacity by production demands.

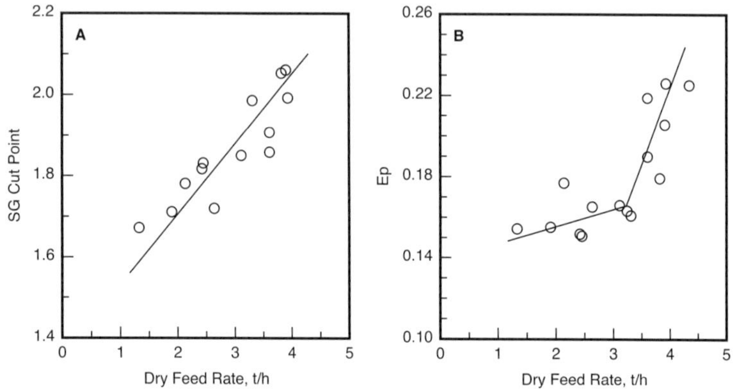

FIGURE 7 Effect of dry solid feed rate on (a) SG cut point and (b) Ep for single-stage spirals

Slurry Flow Rate

Spirals must be provided with an adequate and stable slurry flow rate to work properly. At a constant tonnage of dry solids, a reduction in volumetric flow increases the feed solids content and typically decreases the SG cut point. This action also generally lowers the clean coal recovery and improves the clean coal ash. The optimum slurry flow rate varies in accordance with the spiral diameter (i.e., larger spirals are capable of handling a larger flow rate). Most industrial units need a volumetric flow rate of approximately 7–8 m^3/h (30–35 gpm) per start. Somewhat higher flows would typically be more desirable for treating coarsely distributed material, whereas lower flows are preferred for a more finely distributed material. Because slurry flows are difficult to measure in practice, the proper flow rate for spirals can often be determined by visual inspection of the slurry level up the outer wall of the spiral trough. Most commercial spirals are designed so that proper flow rate occurs when the liquid level comes within about 1.0–1.5 cm (0.4–0.6 in.) from the top of the spiral trough. Too little flow can result in sluggish movement of solids along the interior surface of the spiral that can eventually lead to beaching or sanding. The buildup of coarse particles caused by beaching or sanding can push rock into clean coal or trap clean coal in the refuse stream. On the other hand, too much flow can cause high-density rock to report to the low-density stream, again reducing the quality of the clean product. In most cases, the proper dry solids and volumetric flow rates result in a solids content to the spiral feed of about 25%–35% by weight.

Splitter and Cutter Positions

Spirals are equipped with adjustable splitters located at the end of the last spiral turn (Figure 8). The splitters make it possible to produce three products (i.e., clean coal, middlings, and refuse) from the spiral. The outboard splitter, located near the outer wall of the spiral, divides the clean coal from the middlings. This splitter generally is capable of making a SG cut point between approximately 1.55 and 2.0 SG. The inboard splitter, located closest to the central support stick, divides the middlings from the refuse stream. This splitter typically can cut between 1.8 and 2.4 SG. However, these SG ranges are highly dependent on particle size and washability characteristics and on the feed slurry and dry solids flow rates (Mikhail et al. 1988; King et al. 1992). Some spirals are also equipped with refuse cutters located along the height of the spiral to assist in the removal of rock from high-ash feed coals. The positions of the splitters and cutters generally are selected to fine-tune the separation to the desired target quality. In a typical case, the outboard splitter is located approximately 2.5 cm (1 in.) back from its full outward (outside wall) position, and the inboard splitter is located approximately 5 cm (2 in.) back from its full inward (mounting stick) position. Cutters are typically opened to about 5 cm (2 in.) from full closure or to a position dictated by the refuse loading.

Feed Distribution

All spirals suffer from an inherently low feed capacity. As a result, spirals are installed in banks of large numbers of parallel units to meet typical plant production needs. The maximum yield from such a circuit can be realized only by operating all the spiral units at identical SG cut points. This is true regardless of the desired quality of the total clean coal product or the ratios of different coal seams passed through the spiral circuit (Abbott 1981; Luttrell et al. 2000). To avoid the differences in cut point, spirals must receive a stable supply of feed slurry and be set up identically in terms of splitter and cutter settings. Failure to do so can result in significant losses of clean coal yield. Unfortunately, poor feed distribution is a common problem in industrial spiral banks. Poor distribution can create differences in SG cut points between different spiral units and, in extreme cases, can lead to severe operational problems such as beaching or sanding. Poor distribution can be attributed to variations in slurry flow rates caused by blocked distributor ports or by an incorrect operating level in the feed pot. The slurry level in the distributor pot should be kept at the recommended depth, which is usually 30–45 cm (12–18 in.) to maintain the hydrostatic head necessary to balance the flow rates from the discharge ports.

FIGURE 8 Manually adjustable splitters for fine-tuning coal spiral SG cut points

Particle Size

Coal spirals are typically used for treating coal particles in the 1 × 0.15 mm (14 × 100 mesh) size range. Particles coarser than this upper limit begin to report preferentially to the clean product regardless of density, and finer particles lack sufficient mass to be sorted and tend to report in proportion to the water flow. In fact, many plants equipped with flotation cells have found the lower size limit of about 0.25 mm (65 mesh) to be better because flotation often provides a superior separation for this finest size fraction. In any case, every effort should be made to ensure that the feeding of oversize material to spirals is avoided. In addition, the spiral products must be efficiently sized to remove ultrafine clays that report with the water to the clean coal product. A recent study by Barbee and Nottingham (2007) showed that two stages of fine wire sieves can be used effectively for this purpose.

One interesting development related to particle size is the recent appearance of ultrafine coal spirals. This proprietary circuitry has been designed specifically to upgrade coals that cannot be recovered by froth flotation because of surface oxidation or other extenuating circumstances. In this system, feed coal slurry is treated by multiple classification stages to yield a spiral feed that is nominally 0.15 × 0.044 mm (100 × 325 mesh). The deslimed feed is fed to conventional spirals at a low rate of 0.45–0.50 t/h of dry solids per spiral start and about 12% solids. Under these conditions, ultrafine spirals have been reported to achieve a 1.8 SG cut point and an Ep of 0.20–0.25 (Honaker et al. 2006). However, to be effective, the residual minus-0.044-mm particles in the clean product must be removed by post-classification systems, which is proving to be a challenge in industrial practice. The results obtained to date have been mixed and suggest that the circuitry has the potential to be effective in treating well-liberated low-ash coals with high pyrite contents but may not be suitable for poorer feed coals with a high proportion of middlings or clay.

SUMMARY

The recent implementation of two-stage cleaner spirals has made it possible to efficiently recover fine coal contained in the 1 × 0.15 mm feed to preparation plants at a level once believed to be unlikely for water-based separators. This modern configuration lowers the SG cut point and reduces the bypass of high-ash rock that has been a longstanding problem for traditional single-stage spirals. In addition, the deployment of compound units that include two stages of cleaner separation along a single stick has reduced the complexity and associated costs of implementing this technology in the coal industry. However, optimal performance can be achieved only when these modern circuits are properly operated and maintained. Important parameters that must be monitored include dry tonnage rate, volumetric flow rate, feed distributor levels, splitter and cutter positions, and particle size distributions. When properly designed and operated, spiral circuits provide efficient separations and years of trouble-free service.

ACKNOWLEDGMENTS

The authors acknowledge the continued support by various industrial participants who have contributed to this chapter. In particular, the authors recognize the contributions of the Center for Advanced Separation Technologies, Arch Coal, Alpha Natural Resources, Precision Testing Laboratories, Roche Mining, and PrepTech.

REFERENCES

Abbott, J. 1981. The optimisation of process parameters to maximise the profitability from a three-component blend. In *Proceedings of the First Australian Coal Preparation Conference*, Newcastle, Australia, April 6–10. Edited by A.R. Swanson. Broadmeadow, Australia: Newey and Beath Printers. pp. 87–105.

Barbee, C.J., and Nottingham, J. 2007. Black Bear prep plant replacement of high frequency screens with fine wire sieves. In *Proceedings of the 24th Annual Coal Preparation Conference and Exhibition*, Lexington, KY. New York: Penton Media. pp. 113–122.

Bethell, P.J. 2002. Fine coal cleaning at Massey Energy. In *Proceedings of the 19th Annual Coal Preparation Conference and Exhibition*, Lexington, KY. Stamford, CT: Primedia Business Exhibitions. pp. 1–13.

Bethell, P.J., and Arnold, B.J. 2003. Comparing a two-stage spiral to two stages of spirals for fine coal preparation. In *Advances in Gravity Concentration*. Littleton, CO: SME. pp. 107–114.

Bethell, P.J., Stanley, F.L., and Horton, S. 1991. Benefits associated with two-stage spiral cleaning at McClure preparation plant. Presented at SME Annual Meeting, Denver, CO.

Colorado Historical Society. 2001. Grant-Humphreys Mansion. www.coloradohistory.org. Accessed June 2007.

Holland-Batt, A.B. 1994. The effect of feed rate on the performance of coal spirals. *Coal Prep*. 14:199–222.

Honaker, R.Q., and Wang, D. 1991. *A Spiral Concentrator Study for the Treatment of Fine Coal*. Technical Project Report. Department of Mining Engineering, Southern Illinois University, Carbondale, IL (uncirculated).

Honaker, R.Q., Jain, M., Parekh, B.K., and Saracoglu, M. 2006. Ultrafine coal cleaning using spiral concentrators. In *Proceedings of the XV International Coal Preparation Congress*, October 17–20. Beijing: China University of Mining and Technology. pp. 502–508.

King, R.P., Juckes, A.H., and Stirling, P.A. 1992. A quantitative model for the prediction of fine coal cleaning in a spiral concentrator. *Coal Prep*. 11:51–66.

Li, M., Wood, C.J., and Davis, J.J. 1993. A study of coal washing spirals. *Coal Prep*. 12:117–131.

Luttrell, G.H., McKeon, T., and Kohmuench, J.K. 2007. Design criteria for multistage spiral circuits. In *Proceedings of the SME Annual Meeting and Exhibit*, February 25–28. Preprint 07-126. Denver, CO: SME.

Luttrell, G.H., Kohmuench, J.N., Stanley, F.L., and Trump, G.D. 1998. Improving spiral performance using circuit analysis. *Miner. Metall. Eng*. 15(4):16–21.

Luttrell, G.H., Catarious, D.M., Miller, J.D., and Stanley, F.L. 2000. An evaluation of plantwide control strategies for coal preparation plants. In *Control 2000*. Edited by J.A. Herbst. Littleton, CO: SME. pp. 175–184.

Mikhail, M.W., Salama, A.I.A., Parsons, I.S., and Humeniuk, O.E. 1988. Evaluation and application of spirals and water-only cyclones in cleaning fine coal. *Coal Prep.* 6:53–78.

Osborne, D.G. 1988. *Coal Preparation Technology*, Vol. 1. Boston: Graham and Trotman. pp. 347–365.

Subsinghe, G.K., and Kelly, E.G. 1991. Model of a coal washing spiral. *Coal Prep.* 9:1–11.

———. 1992. Predicting the cut specific gravity of a coal washing spiral. *Miner. Eng.* 5(2):193–203.

Weldon, W. 1997. Fine coal beneficiation: Spiral separators in the Australian industry. *Aust. Coal Rev.* (November):25–28.

Application of Derrick Corporation's Stack Sizer Technology in Clean Coal Spiral Product Circuits

Paul Brodzik

ABSTRACT

This chapter discusses the successful introduction of Derrick Corporation's Stack Sizer technology into the clean coal preparation process. The Stack Sizer is a multideck, high-frequency vibrating screen capable of separations as fine as 75 µm when fitted with Derrick Corporation's patented high open-area urethane screen panels. In 2006, the James River Coal Company selected the Stack Sizer for removal of minus-150-µm (100-mesh) high-ash clay fraction from the clean coal spiral product circuits at the McCoy Elkhorn Bevins Branch preparation plant and at the Blue Diamond Leatherwood preparation plant.

Before installation of the Stack Sizer, sieve bends were used in an attempt to remove the high-ash fraction from the clean coal produced by the spirals. The sieve bends normally produced a clean coal product that was approximately 15%–17% ash. This exceeded the desired ash content of 10% or less.

Full-scale laboratory tests and more than 10 months of continuous production have confirmed that the Stack Sizer fitted with Derrick 100-µm urethane screen panels consistently produces a clean coal fraction which ranges from 8% to 10% ash. Currently, each five-deck Stack Sizer operating at the Bevins Branch and Leatherwood preparation plants is producing approximately 33 short tons per hour of clean coal containing about 9% ash. This represents a clean coal yield of about 75% and an ash reduction of about 11% from the feed slurry.

INTRODUCTION

For more than 30 years, Derrick Corporation's high-frequency vibrating screens have been used by the coal industry in both dry and wet fine-sizing applications. The majority of wet sizing applications have relied on Derrick's multifeed wet sizing machine to make separations as fine as 100 µm in an effort to reduce the high ash fraction from clean coal circuits. The multifeed technology was introduced in the 1970s as a high-capacity wet sizing screen. It consisted of three independent 30-in.-long screen sections that were built into a single vibrating frame—essentially three small screening machines in one unit. Patented Derrick sandwich wire screens were installed as the screen surfaces for many of the coal applications. Each machine was fitted with a Derrick 1.5-hp, 3,600-rpm vibrating motor.

As time went by, coal preparation plants demanded higher-capacity screening machines with longer-lasting screen surfaces. Derrick responded first to the coal industry's needs for longer-lasting screen surfaces by developing high open-area urethane screen panels. Derrick urethane panels achieve capacities comparable to those of wire cloth, but they routinely last 20 to 30 times longer than comparable wire panels. In addition to longer panel life, Derrick urethane screen panels will not blind with near-size particles because of the taper relief built into the panel openings. Wire screen panels in many multifeed machines were replaced with the Derrick urethane panels wherein panel life increased from several weeks to more than 12 months while the processing tonnage remained essentially the same.

In the late 1990s, Derrick Corporation began developing a new concept for high-capacity fine wet screening: the Stack Sizer. The machine design is based on the following wet sizing principles:

- Sufficient water coupled with high-frequency vibration is necessary to allow fine solids to pass through a screen surface. Sizing will stop when all available free liquid has passed through the screen surface.
- Slurry must be introduced to the screen surfaces in uniform, thin layers.
- Oversize material must be quickly conveyed off the screen surface to free up effective screening area.
- Width is the most important factor in determining wet screening capacity and efficiency.

Figure 1 illustrates the final design of Derrick's five-deck Stack Sizer as it is configured for coal applications.

Derrick engineers used the principle of linear vibrating motion to provide consistent motion along the entire length of each screen deck and to all five of the interconnected decks. The motion is achieved by two Derrick 2.5-hp, 1,800-rpm vibrating motors rotating in opposite directions. The high-frequency linear motion is responsible for delivering sufficient force to accelerate the

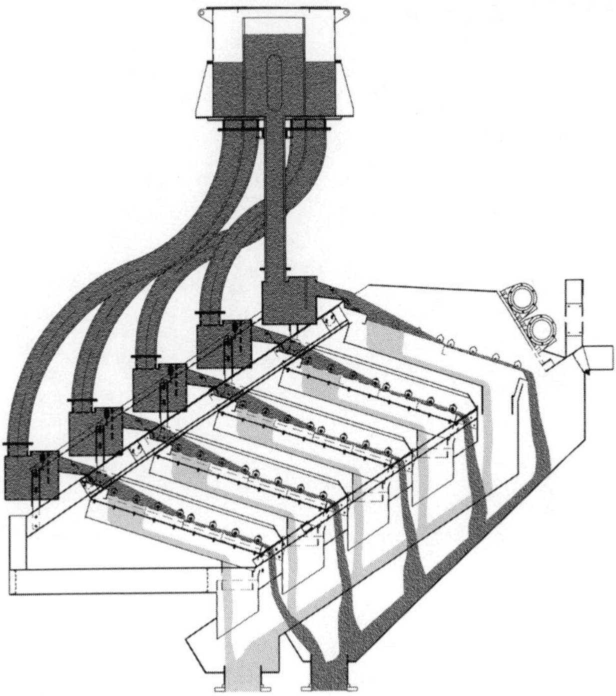

FIGURE 1 Derrick five-deck Stack Sizer (Model 2SG48-60R-5STK)

incoming feed slurry into a thin layer down the entire length of the screen surface. This increases the effective fluidized zone where efficient sizing occurs. The linear motion is also responsible for delivering sufficient force to convey the oversize solids off the screen surface and increase the effective screening area. High capacity is achieved by having up to five 48-in. screen decks operating in parallel on one vibrating frame. The Stack Sizer is essentially five machines in one. It has an effective width of 240 in., constructed in the footprint of a traditional machine having only 48 in. of effective width.

Using the principle of combining sufficient water with high-frequency vibration for efficient removal of fine solids, Derrick engineers incorporated a repulping trough into the design of each screening deck. Each Stack Sizer screen deck consists of two 48-in. × 30-in. screen sections separated by a repulping trough. Addition of free water to the screened oversize fraction provides additional fluid for the removal of trapped fines. Free water is added to the screened oversize fraction as it enters the repulp trough. Water is added countercurrent to the movement of the oversize solids. This creates turbulence in the repulp trough that liberates fine solids from the coarse particles and

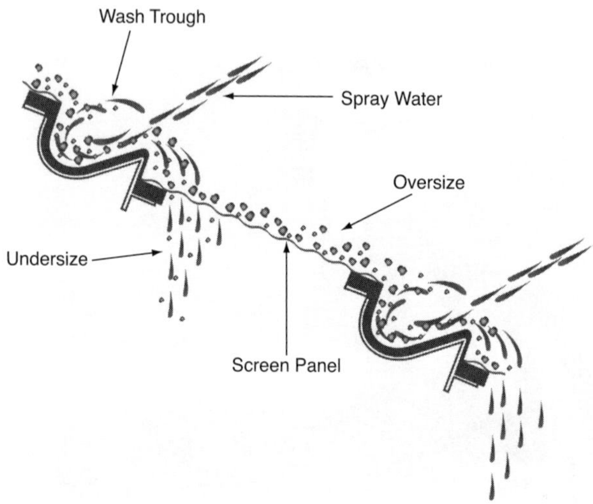

FIGURE 2 Derrick repulp trough

allows them to pass the screen surface with the fluid as it feeds onto the second screen panel. Figure 2 illustrates the action of the repulp trough.

After successful introduction of the Stack Sizer into iron ore, phosphate, and sand applications, Derrick application engineers turned toward the coal industry and its existing installations of Derrick multifeed machines. Full-scale laboratory testing on coal slurry from existing Derrick multifeed users began in 2002. Data indicated that one five-deck Stack Sizer fitted with 180-µm screen panels could process approximately 75 stph (short tons per hour) of feed slurry with 23% –100 mesh to produce approximately 57 stph of clean coal that was 7% –100 mesh. The installed multifeed machine was processing only 20 stph of feed slurry to produce 17 stph clean coal that was 11% –100 mesh.

Additional test work confirmed similar capacity increases and –100 mesh coal reduction when the Stack Sizer was fitted with Derrick 180-, 150-, and minus-100-µm screen panels. Table 1 provides typical laboratory data from wet coal slurry sizing tests on the Stack Sizer.

In 2005, the James River Coal Company sent representative coal slurry samples to Derrick Corporation for a test program intended for evaluating the Stack Sizer's performance for ash reduction in clean coal spiral product circuits. The results were consistent with those of previous test programs with other coal slurries. Duplicate samples of the oversize and undersize fractions were collected for additional ash content analysis. The ash content analysis showed that the clean coal product had been reduced from approximately

TABLE 1 Wet sizing of coal on Derrick Stack Sizer (data for one 48-in. × 60-in. screen deck)

	Feed			Oversize		Undersize	
Screen Panel	Rate, stph	Water, gpm	Cumulative % (+100 mesh)	Rate, stph	Cumulative % (+100 mesh)	Rate, stph	Cumulative % (+100 mesh)
TH* × 0.18MT†	11.9	50	61.43	6.6	95.51	5.3	21.03
TH × 0.15MT‡	10.9	50	61.43	6.9	93.87	4.0	6.23

*TH represents a urethane panel.
†0.18MT represents a 0.18-mm opening with a tapered slot.
‡0.15MT represents a 0.15-mm opening with a tapered slot.

20% to 7% when the Stack Sizer was fitted with Derrick 180-μm urethane panels. It also showed that the ash content was reduced to 9% when the Stack Sizer was fitted with Derrick 150-μm urethane panels. The ash content results showed that the clean coal product from the Stack Sizer was nearly equivalent to the ash content in their heavy-media cyclone circuit. These results convinced the James River Coal Company to select the Stack Sizer for their installations at Bevins Branch and Leatherwood preparation plants.

MCCOY ELKHORN BEVINS BRANCH PREPARATION PLANT CASE STUDY

Full-scale wet screening tests were performed on representative clean coal spiral discharge at Derrick Corporation's laboratory in Buffalo, New York, in July 2005. The test results indicated that one five-deck Stack Sizer fitted with Derrick 150-μm urethane screen panels could process up to 55 stph of feed with 39% –100 mesh when the slurry density was between 25% and 30% solids by weight. Approximately 50 gpm (gallons per minute) of repulp water was added per screen deck. The clean coal product from the Stack Sizer was approximately 35 stph with only 7% –100 mesh. Based on these data, one five-deck Stack Sizer was needed to process the clean coal spiral discharge.

One five-deck repulping Stack Sizer (Model 2SG48-60R-5STK) was installed between December 2005 and January 2006. Figure 3 shows the Bevins Branch Stack Sizer.

Startup of the unit began in February 2006. Performance data from the first few weeks after installation showed that the machine was operating as predicted by the laboratory tests. The data suggested that more good coal could be recovered if a finer separation could be achieved. The 150-μm urethane screen panels were replaced with Derrick 100-μm urethane panels as a field trial. The results from the field trial indicated that the clean coal product yield increased with only a slight increase in the ash content. Table 2 provides performance data from the field trial.

FIGURE 3 Bevins Branch Stack Sizer installation

TABLE 2 Clean coal product from Bevins Branch preparation plant Stack Sizer fitted with 100-μm urethane panels

Stream	% Solids	% Ash
Feed	37.05	12.82
Oversize	49.30	9.86
Undersize	2.96	40.11

Based on Table 2 data, the 100-μm urethane screen panels continue in use on the machine. They have been operating for more than 11 months with no apparent performance degradation.

The Bevins Branch Stack Sizer is processing approximately 40 stph of clean coal spiral product, with about 20% ash. The clean coal product yield is approximately 32.5 stph, with about 10% ash. This material is then fed to screen-bowl centrifuges for additional processing.

BLUE DIAMOND COAL COMPANY'S LEATHERWOOD PREPARATION PLANT CASE STUDY

Test data indicated that two five-deck Stack Sizers were needed to process the anticipated feed flow rate of approximately 80 stph of clean coal spiral discharge. Test data also indicated that the maximum screening efficiency occurred when the feed to the Stack Sizer was approximately 25%–30% solids by weight.

Two Stack Sizers (Model 2SG48-60R-5STK) were installed between January and March 2006. Startup of the machines occurred in April 2006. Figure 4 shows the Leatherwood Stack Sizers.

FIGURE 4 Leatherwood Stack Sizer installation

TABLE 3 Clean coal product from Leatherwood preparation plant Stack Sizers fitted with 100-μm urethane screen panels

Stream	% Solids	% Ash
Feed	22.31	21.23
Oversize	34.65	9.67
Undersize	7.71	61.75

Based on the positive results from the Bevins Branch machine, Derrick 100-μm urethane screen panels were installed on both units. Some dilution water was needed to achieve the desired Stack Sizer feed density of 30% solids or less by weight. The dilution water was added upstream of the flow distributor to the Stack Sizer.

After several weeks of operation, samples of the oversize and undersize fractions from the Stack Sizer were collected and analyzed to confirm the performance of the machine. Table 3 provides performance data from one of the sampling events.

Each Leatherwood Stack Sizer is processing approximately 40 stph of clean coal spiral product, with about 20% ash. The clean coal product yield is approximately 32.5 stph, with about 10% ash. This material is then fed to screen-bowl centrifuges for additional processing. The 100-μm screen panels originally installed on the machines are still in use and show no apparent degradation in performance.

OTHER APPLICATIONS FOR THE COAL INDUSTRY

With the successful application of the Stack Sizer for ash reduction in clean coal spiral circuits, Derrick Corporation has begun evaluation of other applications for the coal industry. Full-scale tests with the Stack Sizer are being conducted to evaluate the machine's performance for ash reduction from 6-in. hydrocyclone underflow and for fine coal recovery from settling ponds.

Recent tests on samples provided by a southern West Virginia coal company indicate that the Stack Sizer fitted with Derrick 75-µm urethane screen panels can effectively reduce the ash content from 6-in. hydrocyclone underflow. One five-deck Stack Sizer can process about 23 stph of feed slurry, with 20% solids by weight. The Stack Sizer reduced the ash content of the clean coal from 37% in the feed to 8% in the oversize fraction.

Other recent tests on samples provided by a Kentucky coal company indicate that the Stack Sizer fitted with Derrick 75-µm urethane screen panels can process approximately 20 stph of slurry (25%–30% solids by weight) dredged from settling ponds. The tests showed a reduction of the high-ash, –75 µm fraction from 58% in the feed to 7% in the oversize fraction.

CONCLUSIONS

The Derrick Stack Sizer has proven its ability to provide efficient, high-capacity fine screening for wet coal sizing applications that require significant ash reduction. Derrick's fine urethane screen panel technology has proven to be a long-lived screen surface for coal applications that require separations as fine as 75 µm. Field data indicate that the 100-µm screen surfaces are lasting longer than 10 months in continuous use.

Current test programs will show the effectiveness of the Stack Sizer for other wet screening applications in the coal preparation industry. Full-scale test programs are essential for proper equipment selection and sizing as the machine's capacity, screening efficiency, and cut points are affected by several factors, including feed percentage solids, percentage oversize, percentage near-size, flow rate, and screen panel tension.

Fine Coal Cleaning: A Review of Column Flotation Options and Design Considerations

J.N. Kohmuench, M.J. Mankosa, and G.H. Luttrell

ABSTRACT

Column flotation has proven to be an efficient and economical means of recovering fine coal. When designed properly, flotation columns provide a high combustible recovery while maintaining a low product ash. This selectivity is the result of a combination of effective sparging and simultaneous froth washing, which eliminates the nonselective hydraulic entrainment of ultrafine ash-bearing minerals to the clean coal product. During the last decade and after the installation of more than 65 coal flotation columns, Eriez Manufacturing continues to gain insight into the proper selection and design of these devices and their associated circuitry. This understanding will be essential as high-grade coal deposits continue to diminish and the processing of high-ash feed stocks increases. A review of the primary column flotation design parameters and other circuit design considerations is presented.

INTRODUCTION

The most widely accepted method for upgrading fine coal is froth flotation. Unfortunately, conventional mechanically agitated flotation machines allow some portion of the ultrafine mineral slimes to be recovered with the water that reports to the froth. This particle entrainment is the nonselective hydraulic conveyance of gangue into the product launder. The liquid phase that holds the froth together and maintains mobility also carries fine particles that have not been collected through a bubble–particle attachment. Fine particles (<0.045 mm) tend to report to the froth concentrate in direct proportion to the amount of product water. Therefore, the flotation operator often is forced to make the decision to either "pull hard" on the cells to maintain yield or run the cells less aggressively to maintain product grade.

Column cells can overcome this shortcoming by rinsing the clay slimes from the froth product using a countercurrent flow of wash water (Figure 1). This feature makes it possible for column cells to achieve superior ash removal when compared with conventional flotation cells. However, sufficient wash water must be added to ensure that all of the feedwater that normally reports to the froth product has been replaced with fresh or clarified water. The froth depth must also be much greater than that used in conventional flotation to obtain good distribution of the wash water and to achieve the desired froth cleaning action.

Unlike in a conventional mechanical cell, feed slurry enters a flotation column and is distributed throughout the cross-section of the separator. The slurry moves countercurrent to a rising swarm of fine bubbles that are generated by an air sparging device. Particles that collide with and attach to the bubbles are carried up the column and discharged to the product launder. Particles that do not attach settle down through the column and report to the underflow tailings stream. In practice, fresh or clarified wash water is showered over or into the froth as it is discharged from the cell. The water filters down through the froth and removes entrained particles and fine clays. The result is an improved purity of the final product. Thus, column flotation cells offer an efficient means to simultaneously reduce the occurrence of entrainment and maintain combustible recovery.

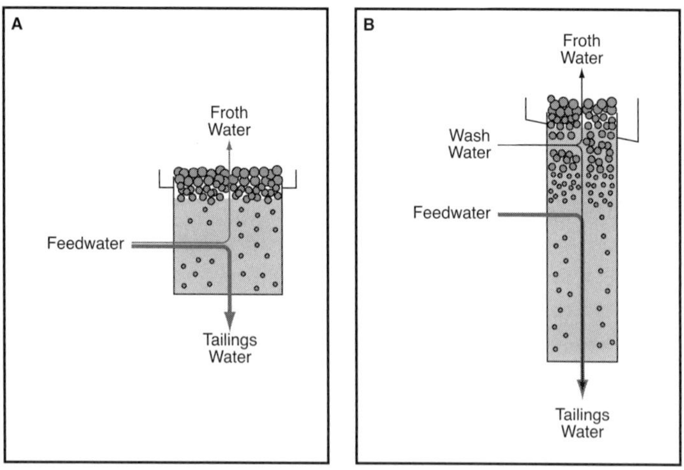

FIGURE 1 Comparison of feed entrainment in (a) mechanical flotation and (b) column flotation machines

The primary advantage of column flotation cells is the ability to provide a superior separation performance compared with conventional flotation processes. This capability is illustrated by the test data summarized in Figure 2, which compares column flotation technology with an existing bank of conventional cells. As shown, the separation data for the column cells are far superior to those obtained from the conventional flotation bank. In fact, the data for the column cells tend to fall just below the separation curve predicted by release analysis (Dell et al. 1972). A release analysis is an indication of the ultimate flotation performance and is often regarded as "washability" for flotation. This figure suggests that columns provide a level of performance that would be difficult to achieve even after multiple stages of cleaning by conventional machines. In addition, the data from both the laboratory (5-cm diameter) test column and full-scale (3-m diameter) column fall along nearly identical recovery grade curves. This finding suggests that the recovery grade response obtained using small test columns can be used to confidently predict the performance of full-scale columns.

The importance of minimizing entrainment is easily understood when the specific gravities (i.e., values) of components in a typical coal stream are

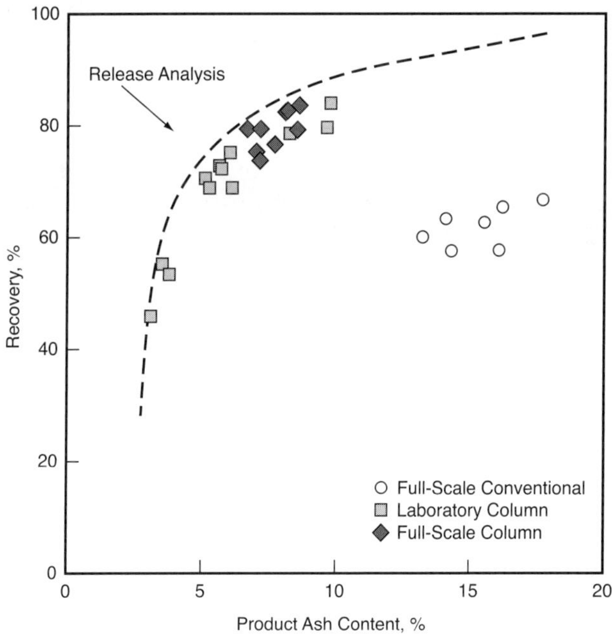

Source: Davis et al. 1995.

FIGURE 2 Comparison of column and conventional flotation cells

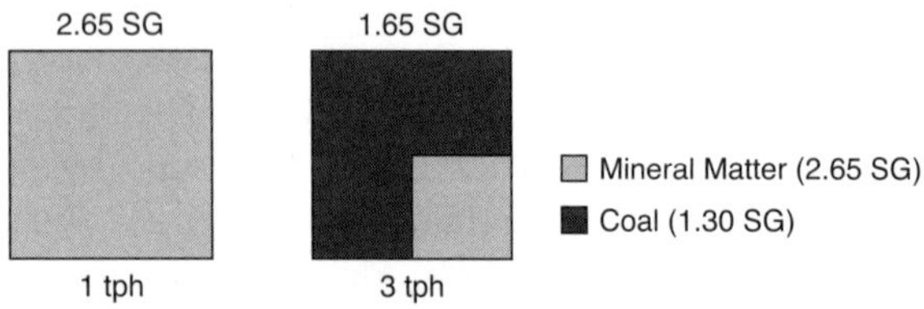

FIGURE 3 Comparison of various coal particles

compared. Simply stated, when a product stream contains more carbon (i.e., lower specific gravity [SG]), the value of that stream is higher than when it consists predominantly of rock or mineral matter. This can be further examined on a particle-by-particle basis, as seen in Figure 3. For example, a particle that has an effective SG of 1.65 consists of 25% ash-bearing minerals by volume. This particle is obviously desired and often ends up as clean coal. In contrast, a particle that is nearly 100% mineral matter has an effective density much greater than 2.0. This particle is not preferred, but a large portion often reports to the product launder of conventional mechanical cells through entrainment. As a consequence, other plant circuits must run at lower specific gravities to maintain the overall plant product grade. In contrast, column flotation can be used to reduce entrainment and allow other plant circuits to run at higher specific gravity cut points. The result is an increase in overall plant yield. Based on a comparison of the specific gravities of these particles, 3 tons of 1.65-SG material can be captured for every ton of clay and mineral matter removed from the flotation product stream while maintaining the same overall product quality.

SPARGING OPTIONS

The ability to consistently maintain this high level of performance is a result of efficient froth washing, which can be achieved only when columns are operated with deep froths (>450 mm). In order to run with deep froths, the sparging systems must be designed to provide the maximum rate of bubble surface area through the column. Therefore, the air sparging system is perhaps the most important component in a column flotation cell. Although details related to the specific design features of the various sparging technologies have been presented in the literature (McKay et al. 1988; Davis et al. 1995; Finch 1995), the two most common column sparging systems offered in U.S. and Australian coal markets are presented here.

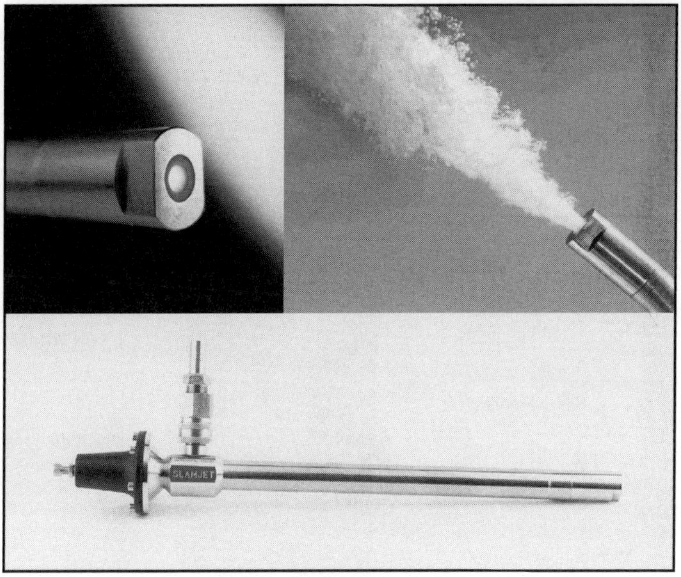

Courtesy of Canadian Process Technologies, Vancouver, Canada.

FIGURE 4 SlamJet sparger with self-closing nozzle

SlamJet Technology

The first practical bubble generation system was developed by the U.S. Bureau of Mines. This design consisted of a series of small orifices (2–3 mm) placed along the length of a pipe, which was then inserted through the column side wall. This approach worked satisfactorily but was plagued with maintenance problems caused by plugging of the orifices. Canadian Process Technologies (CPT) solved the plugging problems with the introduction of their SparJet aeration system. This system uses a series of removable air lances, which includes a single orifice located at the end of the sparger. High-velocity air is injected into the column cell to create and disperse fine bubbles. This technology was further refined when CPT updated the SparJet spargers by incorporating a self-closing mechanism that eliminates backflow of slurry into the aeration system. This updated version is called the SlamJet and is presented in Figure 4. The SlamJet technology has been best applied to 0.150 × 0.045 mm flotation circuits (60M × 0).

Microcel Technology

For minus-0.150-mm (–100M) flotation circuits, one of the most popular sparging technologies used in the coal industry is the Microcel system. This technology was conceived after several years of fundamental research at the

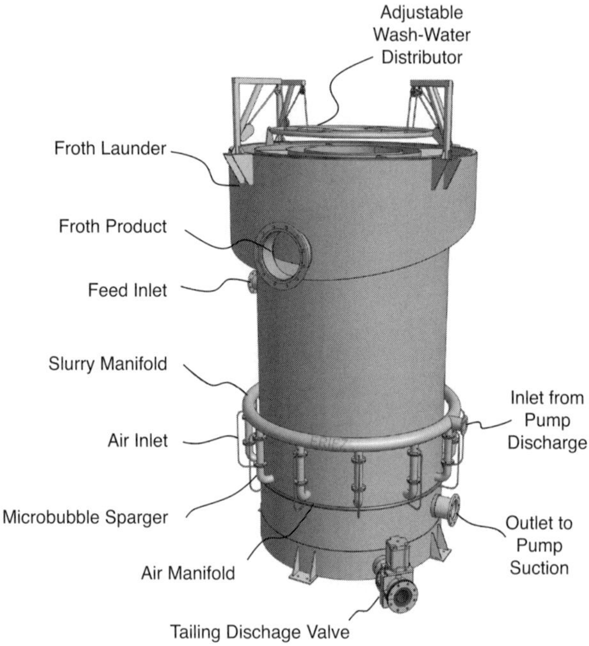

FIGURE 5 Microcel sparging system with essential design features

Virginia Tech Center for Coal and Minerals Processing which showed that the rate of flotation could be enhanced through the use of smaller air bubbles. The small (<0.8 mm) bubbles are generated by circulating slurry from the lower section of the column through parallel in-line static mixers into which compressed air is injected (Figure 5). The Microcel sparging system efficiently generates large amounts of very small bubbles for a given airflow. This capability is commonly reported in terms of the superficial bubble surface area rate (S_b) and is defined as the total bubble surface area per unit of time passing through a given column cross-sectional area. Studies indicate that higher S_b values signify that more bubble surface area is being generated for a given gas rate and are a direct result of a finer bubble size distribution. Therefore, the probability of bubble–particle collisions is greatly increased, which has a direct bearing on capacity. This is an added benefit in treating coal in a traditional minus-0.150-mm circuit, which typically has a high volume flow, shorter residence time, and finer feed size distribution.

COLUMN CELL DESIGN CONSIDERATIONS

In the past decade, Eriez has continued its research and development efforts in circuit design and column sizing. Important design parameters include product carrying capacity, retention time, aeration rate, wash-water rate, and chemical dosage. Tracer studies indicate that mixing is also an important consideration. Proper application of these design parameters is essential to ensure a successful column flotation cell installation.

Carrying Capacity

The product carrying capacity is defined as the mass rate of flotation concentrate per unit time per cell cross-sectional area. Maximum carrying capacity is achieved when all available bubble surface area has been consumed. At this point, any additional coal in the system will be lost to the tailings; therefore, it is important to design column cells with sufficient cross-sectional area to ensure a high combustible recovery. For mineral applications in which the expected float yield is quite low, this goal is rarely attained. However, for coal applications, the float yield can be greater than 80% by weight.

Although product carrying–capacity restrictions can often be ignored for conventional flotation circuits, they are of great importance in the design of column cells. This is largely because the specific surface area of the cell (ratio of the cross-sectional area to the volume) is much higher for conventional cells. Maximum carrying capacity has been well defined by numerous researchers, with extensive data available for coal and mineral applications. These studies indicate that carrying capacity is linearly related to the size and density of froth particles (Sastri 1996). Studies and extensive test work conducted by Eriez also support this finding.

Carrying capacity is best estimated from laboratory and pilot-scale flotation testing. During these evaluations, the feed rate (dry solids) to the column cell is continually increased. Therefore, the product rate increases in direct proportion to the feed rate until the maximum froth capacity is reached, at which time all of the available bubble surface area has been consumed. Up to this point, a linear relationship is realized, with a slope defined as the product yield. Once this maximum froth carrying capacity is reached, the product rate remains constant, as seen in Figure 6. Most full-scale columns in the coal industry typically operate at carrying capacities in the range of 0.9–3.0 $t/h/m^2$ (0.09–0.30 tph/ft^2), with an average of about 1.2 $t/h/m^2$ (0.12 tph/ft^2) for –0.150 mm (100M × 0) feeds. The values at the lower end of this range typically correspond to finer circuits, and values at the higher end correspond to coarser feeds. When the carrying capacity is known, the necessary column cross-sectional area can be determined by dividing the expected clean coal tonnage (t/h) by the carrying capacity ($t/h/m^2$). However, it is generally more

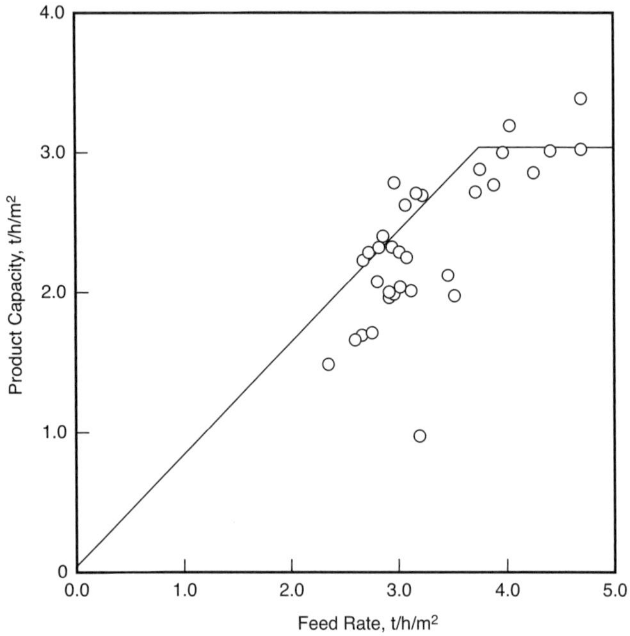

Source: Kohmuench and Mankosa 2006.

FIGURE 6 Typical carrying-capacity plot for pilot-scale test work

convenient to calculate the product tonnage for a full-scale column from a smaller test unit using

$$\frac{\text{large column (t/h)}}{\text{small column (t/h)}} = \left(\frac{\text{large diameter}}{\text{small diameter}}\right)^2 \quad \text{(EQ 1)}$$

Retention Time

Like all flotation equipment, column flotation cells must be designed to provide sufficient slurry residence time to achieve the target recovery of clean coal. The mean residence time of the separator can be estimated by dividing the active volume of the flotation pulp, V, by the volumetric flow rate of refuse, Q_s, as seen in Equation 2.

$$\tau = V/Q_s \quad \text{(EQ 2)}$$

A detailed procedure for estimating the column residence time needed to achieve a given recovery has been presented in the literature (Dobby and Finch 1986) and is beyond the scope of this chapter. However, as a rough guideline, a column cell typically requires at least twice the residence time of a bank of conventional cells and three times the residence time of a batch laboratory flotation cell. Therefore, it would not be unusual for a column cell to require 7–10 minutes of residence time. Furthermore, some industrial columns have been designed to operate with residence times of 12–15 minutes or more. It is also important to note that the active volume for columns is defined as the volume of pulp contained between the bottom of the froth and the top of the air spargers. This volume is substantially less than the total volume of the tank structure. The calculated volume must also be reduced by 10%–20% to account for the volume occupied by the air bubbles (i.e., air fraction).

Aeration Rate

The primary advantage of an efficient and commercially available sparging system is the ability to generate large amounts of very small bubbles. As previously mentioned, this capability is commonly reported in terms of the superficial bubble surface area rate (S_b). This value can be calculated as

$$S_b\{1/s\} = 1.67\left(\frac{\text{superficial air rate}\{m^3/h/m^2\}}{\text{bubble diameter}\{mm\}}\right) \quad \text{(EQ 3)}$$

The impact of S_b on flotation recovery is illustrated by the test results in Figure 7 (Kohmuench et al. 2004). These data show that recovery increases sharply as S_b increases above 50/s. and eventually reaches a plateau at 100–150/s. These values indicate that nearly 100,000 m^2 of bubble surface area must pass through a 4.5-m (15-ft) diameter column every minute to maximize combustible recovery. The data also indicate that the relationship between recovery and S_b is generally independent of sparger type. The problem is that many commercial sparging systems simply cannot produce high S_b values. Field tests also suggest that operation at S_b values above 150/s often produces poor separations because of runaway froths that cannot be effectively washed. A superficial gas rate in the range of 70–95 $m^3/h/m^2$ would be suitable for most coal applications. These values correspond to total aeration rates of approximately 990–1,350 m^3/h (582–795 ft^3/min) of air for a full-scale 4.25-m-diameter column. The gas rates at the lower end of the range would generally be used for spargers that generate smaller bubbles, and the higher gas rates are typically needed for less efficient spargers. A proper combination of gas rate and bubble size generally provides a gas holdup in the flotation pulp above 15%. Values above 20% are not unusual for well-tuned circuits.

Caution should be used during the metering of gas flow rates. A properly designed system should be equipped with a flow meter that is calibrated to read correctly at a specified operating pressure. The operating pressure should be held constant through the use of a pressure regulator ahead of the flow

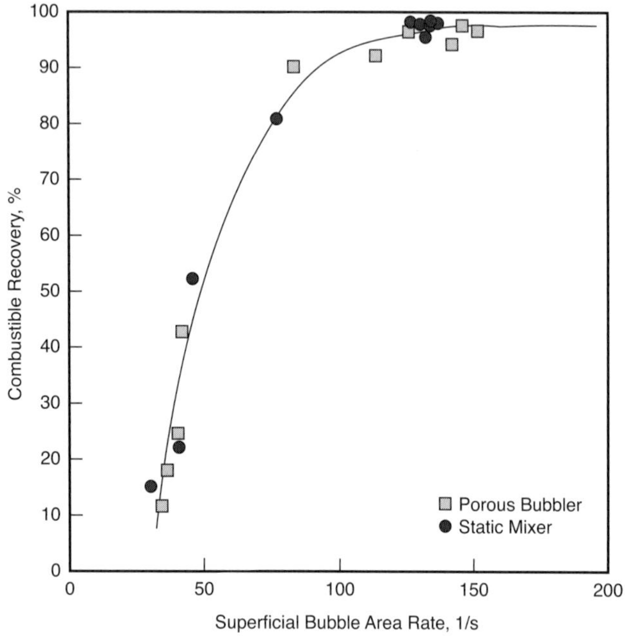

FIGURE 7 Effect of S_b on flotation recovery for different air sparging systems

meter so that the meter always operates at its design pressure. If the flow meter is placed after the control valve, then the operating pressure is unknown and the true gas flow rate cannot be determined. Improper metering of the gas flow rate can be a particularly serious problem when laboratory and pilot-scale tests are conducted for the purpose of collecting scale-up information.

A great deal of confusion also exists regarding the specification of compressors for column applications. Much of this confusion is related to improper use of gas flow terminology (Sullair Corp. 1992). For example, column manufacturers normally report gas flow rates as a standard volumetric flow per time. This value is valid only at 1 atm of pressure and 20°C (68°F) of dry air. The "actual" flow rate specified by compressor manufacturers is typically reported in terms of inlet conditions, or "free air." Although this amount of air enters the compressor, it is not necessarily the amount of air delivered to the column because of compressor seal leakage. As a result, the actual flow may be only 95% of the inlet flow. Furthermore, corrections to the gas flow rate must be made to account for differences in elevation (atmospheric pressure) and humidity. Air temperature generally has little impact on the capacity of an oil-flooded screw compressor but may affect the performance of an air-cooled compressor. These complications generally require that professionals be consulted to ensure that the compressor is properly sized for the specified air requirements.

Froth Washing

The use of wash water in column flotation is necessary to provide the optimum separation efficiency as indicated by release analysis. In addition, a froth depth of 0.6–1.2 m is typically needed to ensure good distribution of the wash water and to prevent short-circuiting. As in any column, the flow of wash water must exceed the volumetric flow of water reporting to the clean coal product to prevent entrainment of the high-ash slimes. In most cases, less than about 1% of the feedwater reports to the froth product if the wash water is properly controlled. The amount of water carried by the froth can be calculated as

$$\text{water demand}\{m^3/h/m^2\} = \text{carrying capacity}\{t/h/m^2\}\left(\frac{100}{\text{froth \%solids}} - 1\right) \quad \text{(EQ 4)}$$

For example, a column cell producing 1.2 $t/h/m^2$ of dry clean coal at 18% solids will carry about 5.5 $m^3/h/m^2$ of water from the pulp into the froth (i.e., 1.2 (100/18 – 1) = 5.5). Entrainment theoretically should be eliminated when the number of dilution washes (defined as the froth-water demand divided by the wash-water addition rate) reaches a value of 1. However, as shown in Figure 8, froth mixing usually requires that 1.25–1.50 dilution washes be used to fully suppress hydraulic entrainment. This constraint dictates that a wash-water flow rate of about 8.3 $m^3/h/m^2$ (i.e., 1.5 × 5.5 = 8.3) be used in the current example to prevent the entrainment of high-ash slimes. Field data collected from columns operating in the coal industry suggest that a wash-water flow rate of 7.3–12.2 $m^3/h/m^2$ (3.0–5.0 gpm/ft^2) is normally adequate for most commercial installations. However, higher gas and frother addition rates will typically increase the froth-water demand and, as a result, the amount of wash water needed. Excessive wash-water flows should be avoided because extra wash water passing downward through a column creates an undesirable reduction in the slurry retention time and hence a potential reduction in recovery. Very high water additions may also destabilize the froth by stripping surfactant (frother) from the bubble surfaces. High water rates may also decrease product grade by increasing axial froth mixing, thereby reducing the wash-water effectiveness (Yianatos et al. 1988).

The design of the wash-water distributor can also significantly affect column performance. In some cases, the distribution piping is intentionally

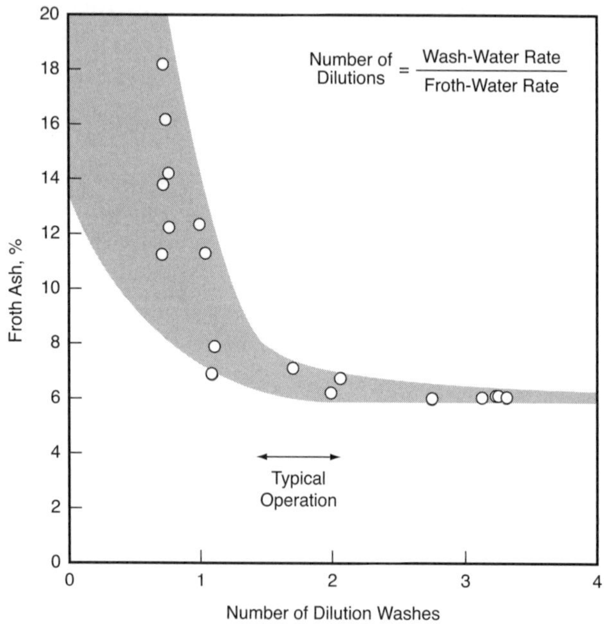

FIGURE 8 Effect of number of dilution washes on clean coal quality

submerged below the cell lip so that a drained froth can form above the distributor. This arrangement allows the depth of the drained froth and the extent of froth drainage to be varied as the distributor is raised or lowered. Changes to the vertical position of the distributor can be used to control the split of water between the clean coal and refuse streams. In some cases, multilevel concentric distribution rings may also be used to overcome problems associated with poor froth mobility. The inner rings are typically located above the outer rings to reduce drainage and improve the fluidity of the froth in the center of the column. More recently, one or more internal launders have been used in lieu of tiered wash-water rings. And in some cases, the water distributor may be located just above the top of the froth. This arrangement does not allow the froth mobility to be controlled by adjustment of the distributor location, but it does make it easier to identify and correct plugging problems that may severely reduce the performance of the distribution network.

Frother Rates

The addition of a surfactant (i.e., frother) is typically needed in most conventional flotation applications. Column flotation is no different and also requires the addition of a frother that will lower the surface tension of the pulp

TABLE 1 Typical frother dosages based on open-circuit designs

Frother Type	Blend, %	Typical Dosage, ppm
Glycol	100	8–10
Glycol/alcohol	70/30	12–14
Glycol/alcohol	90/10	14–16
Alcohol	100	14–18
Methyl isobutyl carbinol (MIBC)	100	20–24

enough to allow the creation of a deep and stable froth. Many types and blends of frothers are available; however, the typical dosage of each type depends on many factors that include frother strength, site water chemistry, flotation circuit configuration, dewatering techniques, and whether the plant maintains an open or closed water circuit. Regardless, sufficient frother must be added to allow maximum flotation yield. The ultimate chemical dosage may be lower if the reagent has the potential to cause other problems in the plant (e.g., froth buildup). Therefore, the choice and dosage of these chemicals should be determined through consultation with professional chemical vendors. Table 1 provides dosage guidelines for various types of frothers. The dosage rates are based on total cell supply volume flows, including wash-water addition.

CIRCUIT DESIGN CONSIDERATIONS

By-Zero and Deslime Circuits

Two typical circuits are used for fine coal flotation: the traditional minus-0.150-mm (–100M) by-zero circuit and the 0.150 × 0.045 mm (100 × 325M) deslime circuit. In the by-zero approach (Figure 9), minus-1-mm feed is sent to large-diameter classifying cyclones. Typically, these cyclones are configured to make a cut point of approximately 0.150 mm (100M). In some markets, this cut point can exceed 0.5 mm (32M), although it is not recommended to exceed 0.250 mm. Regardless of the cut, the cyclone overflow stream is sent directly to flotation.

The second type of circuit that has been gaining popularity, especially for producing coal for the thermal (i.e., steam) market, is the deslime circuit. In this circuit (Figure 10), a secondary bank of 150-mm (6-in.) cyclones is used to further classify the flotation feed at approximately 0.045 mm (325M) for the purpose of rejecting a large portion of the ultrafine clay or coal particles. This approach can be advantageous when the feedstock contains little combustible material in the finest size classes.

Each circuit has certain advantages. The traditional by-zero circuit will always provide the maximum product tons. However, the froth produced from these circuits is extremely stable and typically more difficult to handle. In

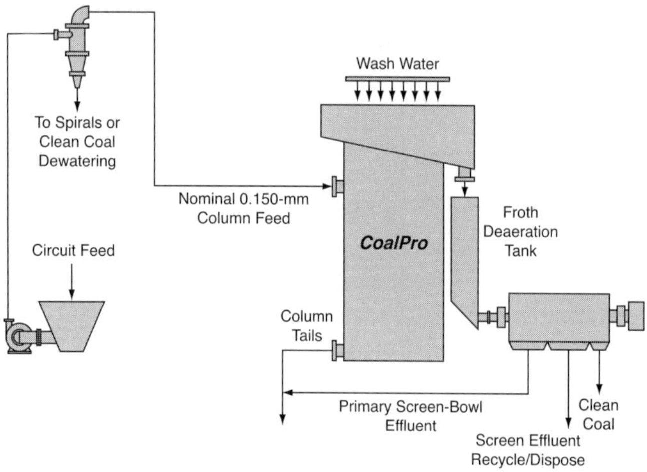

FIGURE 9 Traditional by-zero circuit

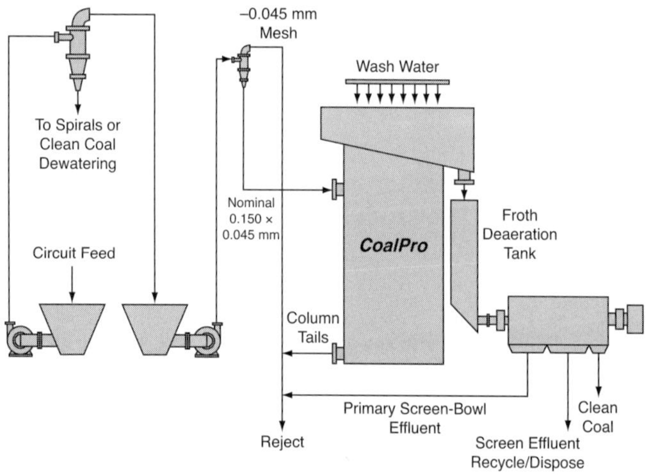

FIGURE 10 Deslime circuit with additional classification

addition, a dewatering scheme that can attain 100% capture (i.e., vacuum filter) should be used to maximize the benefit of the additional product tons. Unfortunately, disk filters and similar vacuum equipment do not obtain the lowest moisture values when compared with screen-bowl centrifuges. These traditional circuits also use more frothing chemical because of the greater volume that must be treated. Because most of this slurry volume is returned to the plant after clarification in a thickener, residual frother can accumulate in the process water and upset other operations in the plant.

On the other hand, the deslime circuit has continued to gain popularity because of its simplicity and ease of operation. The removal of the ultrafine material results in a higher flotation capacity and typically reduces the number or size of columns needed for an application when compared with the traditional circuit. In addition, the froth concentrate is typically coarser and provides a lower product moisture from the associated dewatering equipment. Direct fine coal losses are seen in the cyclone overflow; however, the 150-mm (6-in.) cyclones help to alleviate any problems caused by the concentration of residual frother. In effect, the volume that must be treated with chemical is much lower.

Parallel Versus In-Series Circuitry

As mentioned previously, if sufficient residence time is not provided in the flotation circuitry, coal can be lost to the tailings stream through bypass. Bypass is caused by the internal mixing found in the large-diameter short tanks commonly found in coal installations. The high degree of mixing allows some material in the cell to report to tailings without being influenced by the bubble swarm produced by the sparging system. Typically, as the residence time increases, the probability of bypass decreases. As a consequence of this relationship, the occurrence of bypass is more often found in traditional by-zero circuits where the volumetric feed rates are high and retention times typically are low when compared with those of deslime circuits.

Bypass often is not a problem in conventional flotation cells because of their cell-to-cell arrangement. In contrast, column cells are usually arranged in parallel (Figure 11a). However, recent studies show that arranging column cells in series (Figure 11b) is more beneficial (Stanley et al. 2006). In both approaches depicted in Figure 11, the total residence time for a complete circuit is identical. In addition, the circuit carrying capacity is also equal because the cell diameters (i.e., total cell area) have not changed.

When cells are arranged in series, however, the theoretical maximum combustible recovery as defined by Levenspiel (1972) is significantly greater than when they are arranged in parallel. This can be seen in Figure 12, which shows the expected recovery for a plug-flow system (i.e., infinite number of units in series) and both parallel and in-series circuitry. Specifically, combustible recovery is predicted with respect to a combination of both residence time (τ) and the flotation rate constant (k). Shown are data from tracer studies, which indicate that when operated in parallel, a single column cell achieves a recovery response curve similar to 1.6 mixers in series. A mixer is defined as one perfectly mixed cell. However, this figure also shows that when arranged

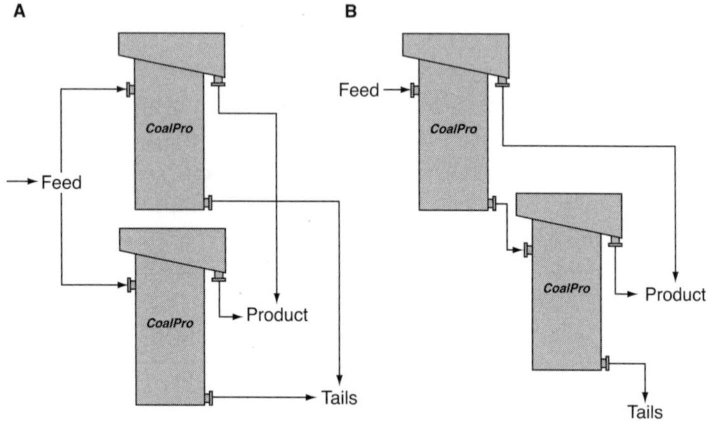

FIGURE 11 Arrangement of cells (a) in parallel versus (b) in series

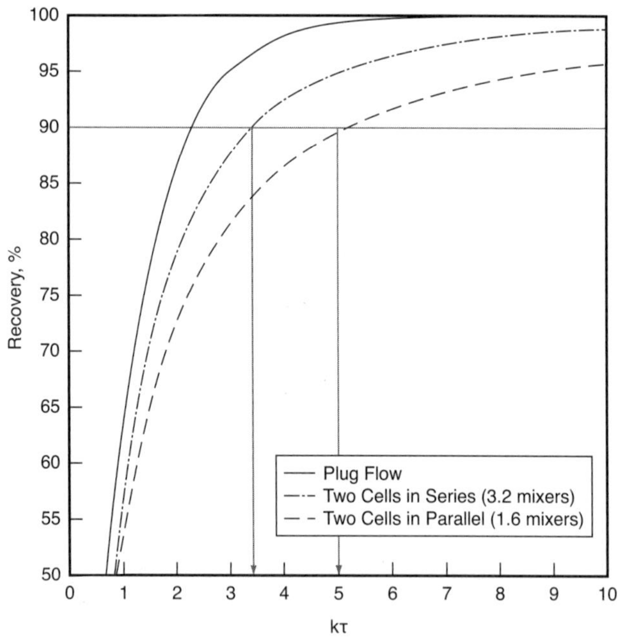

FIGURE 12 Theoretical recovery as a function of residence time and flotation rate constant

in series, a column circuit can achieve a response curve similar to 3.2 mixers in series. In other words, for the same residence time, float capacity, and capital expense, the expected combustible recovery can be greatly increased for any given residence time. Conversely, the time at which a target combustible recovery can be achieved is significantly lower. For instance, in order to reach a combustible recovery of 90%, two columns operating in parallel must exceed a $k\tau$ value of 5.0. However, two columns operating in series achieve this same target recovery at a $k\tau$ value of 3.5. Because the flotation rate constant is equal for a specific coal or system, this reduction indicates that the same combustible recovery can be achieved in 30% less residence time.

Froth Handling

Proper handling of the product can be a problem in column flotation cells because of the large volumes of froth generated by this technology. In particular, concentrates containing large amounts of ultrafine (minus-0.045-mm) coal can become excessively stable, creating serious operating problems related to backup in launders and other downstream unit operations. Attempts to overcome this problem by selecting weaker frothers or reducing frother dosage have not been successful and have generally led to lower column recoveries. Therefore, several circuit modifications have been developed to respond to the froth stability problem. For example, column launders must be greatly oversized, with steep slopes to reduce backup. Horizontal froth travel distances must be kept as short as possible, and adequate vertical head must be provided between downstream operations and column launders.

Some installations have resorted to using defoaming agents or high-pressure launder sprays to cope with the stability problem (Figure 13a). However, newer column installations avoid this problem by including a deaeration tank to permit time for the froth to collapse (Figure 13b). Special provisions may also be needed to ensure that downstream dewatering units can accept the large froth volumes. For example, standard screen-bowl centrifuges equipped with 10-cm (4-in.) inlets may need to be retrofitted with 20-cm (8-in.) or larger inlets to minimize flow restrictions. For very coarse feeds, the froth product can be pumped to classifying cyclones. In such cases, the cyclone underflow (oversize) product is typically passed directly to the dewatering circuit (usually filters), while the cyclone overflow (undersize) product is passed to a clean coal thickener. The solids from the thickener underflow are pumped to the dewatering circuit. The clarified overflow from the thickener is sent back to the column circuit as wash water. This approach maintains a high frother concentration in the column circuit and minimizes the impacts of frother buildup on other plant circuits.

FIGURE 13 Techniques used to combat froth handling problems: (a) high-pressure sprays and defoaming agents and (b) open-top deaeration tank

SUMMARY

As high-grade feedstocks continue to decline, the role of column flotation is likely to increase in future plant designs. Therefore, it is important to understand the engineering and design considerations with regard to this technology. Although the installation of a column circuit can provide a beneficial financial return, the design and scale-up of this technology are challenging. Careful engineering, extensive testing, and attention to ancillary equipment needs will ensure that the column circuit performance agrees with that predicted by standard release analysis.

By itself, proper design of a flotation column includes the understanding of application-specific details, such as coal quality and expected product mass yield. Through this understanding, the cell can be properly engineered with regard to sparging technology, product carrying capacity, retention time, wash-water rates, and aeration rates. Adherence to the column design criteria as outlined in this chapter provides the basis for a successful flotation circuit.

To ensure the success of the entire installation, the impact of the flotation circuitry must also be considered. This includes the determination of the circuit type (i.e., by-zero or deslime) and whether or not the cells should be placed in parallel or in series. Experience has shown that the final market destination (i.e., metallurgical vs. thermal) and contract specifications (i.e., moisture, tons, and inerts) are the primary factors in these decisions. Regardless, outside issues such as froth handling and dewatering techniques must be fully weighed to ensure the proper approach.

Column flotation cells can recover fine coal without contributing fine clay (>90% ash) to the final clean coal product. In this approach, an incremental fraction containing 30%–40% ash can be added to the final product, as opposed to a pure clay component. In general, rejecting 1 tph of fine clay results in a 3-tph increase in plant product tonnage while the plant is operated at the same overall ash content.

REFERENCES

Davis, V.L., Bethell, P.J., Stanley, F.L., and Luttrell, G.H. 1995. Plant practices in fine coal column flotation. In *High Efficiency Coal Preparation: An International Symposium*. Edited by S.K. Kawatra. Littleton, CO: SME. pp. 237–246.

Dell, C.C., Bunyard, M.J., Rickelton, W.A., and Young, P.A. 1972. Release analysis: A comparison of techniques. *Trans. Inst. Min. Metall.* Sect. C 81:787.

Dobby, G.S., and Finch, J.A. 1986. Flotation column scale-up and modeling. *CIM Bull.* 79(889):89–96.

Finch, J.A., 1995. Column flotation—A selected review. Part IV: Novel flotation devices. *Miner. Eng.* 8(6):587–602.

Kohmuench, J.N., and Mankosa, M.J. 2006. An update on the role of column flotation in the coal industry. *Coal Prep. Soc. Am. J.* 5(1):19–26.

Kohmuench, J.N., Davy, M.S., Ingram, W.S., Brake, I.R., and Luttrell, G.H. 2004. Benefits of column flotation using the Eriez Microcel. In *Proceedings of the 10th Australian Coal Preparation Conference*, Polkolbin, New South Wales, Australia, October 17–21. New South Wales: Australian Coal Preparation Society. pp. 272–284.

Levenspiel, O. 1972. *Chemical Reaction Engineering*. New York: Wiley.

McKay, J.D., Foot, D.G., and Shirts, M.D. 1988. Column flotation and bubble generation studies at the Bureau of Mines. In *Column Flotation '88*. Littleton, CO: SME. pp. 173–186.

Sastri, S.R.S. 1996. Technical note: Carrying capacity in flotation columns. *Miner. Eng.* 9(4):465–468.

Stanley, F., King, P., Horton, S., Kennedy, D., McGough, K., and Luttrell, G. 2006. Improvements in flotation column recovery using cell-to-cell circuitry. In *Proceedings of the 23rd International Coal Preparation Exhibition and Conference*, Lexington, KY, May 1–4. Stamford, CT: Primedia Business Exhibitions.

Sullair Corporation. 1992. Industrial Compressors—Compressor Capacity. Sales Bulletin E-78 (April). Michigan City, IN: Sullair Corp.

Yianatos, J.B., Finch, J.A., and LaPlante, A.R. 1988. Selectivity in column flotation froths. *Int. J. Miner. Process.* 23:279–292.

PART 3

How We Can Do It Better

Three-Dimensional Design Technology

Thomas J. Porterfield and Todd Vander Hill

ABSTRACT

This chapter explores the engineering methods at Farnham & Pfile Engineering (F&P) for use in the design-and-build practices of coal preparation plant design. CATIA V5 is a three-dimensional (3-D) database-driven program that relies on the concept of designing systems and structures in an intelligent object-oriented process. This 3-D intelligence allows the engineer to perform database queries that deliver benefits such as parametric models, plant layouts, 3-D attributed models and catalogs, interference analysis, and virtual 3-D navigation.

At the completion of each project, F&P delivers an electronic format for linking engineering and maintenance documents to the 3-D or two-dimensional plant components. Engineering Document Maintenance Document Retrieval is a predefined process using a Web-based-enabled interface to graphically display the plant components that require engineering and maintenance tracking. This graphical interface allows the users to visualize a computer-generated image of the plant from any given computer-aided design system. From this image, the digital plant maintenance provides the tools for retrieving engineering and maintenance documentation. With these tools accessible to every F&P engineer, the design-to-build process has been streamlined to produce a cost-effective and accurate method that sets the standard for future projects.

INTRODUCTION

Coal preparation plant design has moved from the era of drafting boards and entered the three-dimensional (3-D) computer-aided interactive design era. The future coal preparation plant design demands the use of a 3-D computer-aided interactive system to meet all critical needs for an on-schedule, on-budget, and economically operated plant.

Farnham & Pfile Engineering (F&P) has embraced an approach as a benchmark for all projects by using IBM/Dassault Systèmes' CATIA V5 (Computer-Aided Three-Dimensional Interactive Application) and EDMDR (Engineering Document Maintenance Document Retrieval) to design and build future coal preparation plants. In the past 4 years, this approach has helped F&P design and build a 2,000-tph plant, a 1,500-tph addition, and most recently a 2,200-tph plant.

CATIA V5 DESCRIPTION

CATIA V5 is a database-driven program for designing systems and structures in an object-oriented design package, which recognizes every element as what it actually is, not just as lines—from piping to structures to equipment. This is because in digital design, nothing is drawn. A parametric model is a file that represents knowledge engineering. In other words, not only does it have the ability to present a very detailed image, but the model defines specific features such as shape, area, and volume needs, orientation, and inclination and can include other features that may define the element. Parametric models place constraints on the geometry of the image they define, allowing the unit to be resized based on design or application criteria.

Model creation begins with the layout of a grid representing the plant's footprint. The grid, defining the column lines, can be defined by CATIA V5, drawn conventionally in the system, or imported from a file developed in any other drafting program. After this is done, the drawing is over. Elevation planes are established in a 3-D environment to define top of steel at all floor levels. Most of the design is then done at these elevations as though in a two-dimensional (2-D) environment. Although it sounds complex, it is much simpler than using any other computer-aided design (CAD) system.

CATIA V5 is operated primarily through menus. With the model established, dialog boxes allow the user to select view planes or elevations at which to work. Elements representing equipment, plate work, or structural members are then placed at each elevation and can be easily manipulated to give the desired arrangement. Structural members can be arbitrarily placed and then edited later, or the user can elect to work within the model, retaining a stick frame, and insert the actual sections after they are sized. After the structural steel has been designed, detailing can be done directly from the model.

A time-effective application in CATIA V5 is the piping and equipment program. This routine incorporates a rule catalog to determine the type of pipe, size, and routing. Again the engineer needs to develop a library for various types of pipe, pipe sizes, fittings, types of equipment, and so forth. Piping and equipment for each circuit are developed within the plant model and stored in a database. However, the actual piping layout can be pulled separately and displayed, rotated for viewing, and plotted. Like structural steel, piping is sized conventionally, but the process cuts down the time needed to establish and define the piping runs.

Another big advantage of digital design using CATIA V5 is a feature called interference analysis. As each 3-D model is examined in the interference routine, it generates a report when another element invades its space. So if an engineer has routed a pipe through a chute or too close to a beam, the routing error is called out and can be corrected before the pipe is detailed and purchased and, most importantly, before it is delivered to the field. The same procedure can be used with the structural steel, equipment, and plate work, resulting in delivery of nearly perfect materials to the field.

Finally, CATIA V5 gives F&P the ability to see the plant develop in the 3-D model and navigate throughout the model. By navigating through the plant layout, CATIA V5 gives the engineer the tools to walk, fly, and examine each circuit or elevation in all design phases. This enables the designer and owner to work together, ensuring an optimized design for plant layout, maintenance, and possible future expansions. Improvements and corrections can easily be made before they become expensive and before schedules become delayed because of revisions during fabrication and construction.

CATIA V5 provides the following features:

- Predetermined specifications; every object is assigned a logical unit number that remains throughout its life cycle.
- Interference detection during design, nearly eliminating construction interference.
- Virtual 3-D walk-through of the plant, allowing designers, plant owners, and project managers to evaluate all phases of the plant, operations, maintenance, access, and safety.
- Accurate general arrangement of equipment and structural, plate work, and piping drawings, including process and instrumentation diagrams, fabrication, and isometrics.
- Detailed material takeoffs of structural, plate work, and piping layouts.

CATIA V5 provides a set of tools that allows simultaneous design and integration of fluid and mechanical systems in a 3-D digital mock-up while optimizing space allocation. It includes products for optimizing plant layouts, creating circuits, and designing structural products.

EDMDR DESCRIPTION

Whereas F&P uses CATIA V5 3-D design software in every aspect of design development, from conception to final design, EDMDR provides electronic tools for linking engineering and maintenance documents to a 3-D or 2-D computer-generated image of the plant components. These documents can include technical drawings, engineering criteria, supplier documentation, bills of materials, and procedures. EDMDR is a predefined process using a Web-based interface to graphically display the plant components that need engineering and maintenance tracking. This graphic interface allows the users to see a computer-generated image of the plant from any given CAD system. From this image, the digital plant maintenance will provide the tools for retrieving engineering and maintenance documentation.

The concept of EDMDR was developed after years of experience working with and supporting process plant and power industries. This industry experience identified a need to streamline recurring engineering and manufacturing changes that affect decision making about nonconforming material, plant or facility configuration, equipment maintenance, fabrication, and assembly procedures. To make these decisions, technical information, history, status, material, supplier, assembly, and installation requirements must be identified as quickly as possible to support efficient productivity. This means that many drawings, documents, procedures, and documentation must be retrieved and reviewed by the appropriate employees. It can be very time consuming and costly to request drawings, work orders, material disposition, and maintenance documents in many types of formats (e.g., hard copy, word processing, spreadsheets, database, and raster), all of which were not accessible from a common work area, facility, or computer system. The necessary decisions could finally be made but not without many labor hours of document retrieval, correcting inaccurate information, calling, and e-mailing. With EDMDR, F&P has nearly eliminated these problems by providing a solution for all maintenance and document retrieval needs during all phases of plant fabrication.

At the completion of each project, EDMDR also furnishes a complete electronic maintenance manual organized by plant equipment circuits and equipment unit numbers. The owner can print any part of the electronic manual at any time from a desktop or laptop computer. This makes the hard-copy multivolume manuals nearly obsolete.

How does F&P plan to continue using EDMDR for future coal preparation and power plants? F&P can also use EDMDR to create maintenance events. These maintenance events will keep historical records or notify the owner of all scheduled or upcoming maintenance, including dates, labor hours, part documentation, and equipment component life. With notification of upcoming maintenance events, the owner can properly schedule labor hours and ordering or restocking of spare parts, ensuring the best options for cost reduction.

EDMDR also can attach safety and how-to videos for employee viewing to ensure that proper installation procedures are considered during scheduled or nonscheduled maintenance. This allows the plant owner to optimize operations and comply with safety laws and regulations.

EDMDR provides the following features:

- Reduction in the time needed to retrieve information from a given file system or archival location.
- Digital application to determine the personnel needed to support and maintain the plant and its components.
- A hyperlinked Web interface that behaves like Web pages. As a result, new users face almost no learning curve; anyone who has used the Web can readily adapt to EDMDR's streamlined, intuitive interface.
- Customization to suit the specific needs of any project or owner.
- On-line access to plant documentation and history to anticipate and plan for upcoming maintenance.

CONCLUSION

With each project, the quality and experience of F&P grows, improving all aspects of the coal preparation plant. The power of CATIA V5 and EDMDR means that the plant owner gets a coal preparation plant that surpasses the goals for production, safety, and ease of maintenance while remaining on budget and on schedule.

Access for Sampling

B.J. Arnold

ABSTRACT

This chapter identifies several common errors in sample collection at coal preparation plants and discusses proper methods for collecting those samples, including designing plants with access for sampling. The most common errors concerning cyclones and froth flotation circuits are highlighted.

INTRODUCTION

The use of proper sampling methods makes life much easier for the coal preparation engineer, plant operator, plant designer, and equipment supplier. Attention to small details such as using the correct sampling device or tool, not letting containers overflow, and marking sample containers properly reduces sampling, laboratory analysis, and data reduction costs because procedures are performed correctly the first time.

This chapter provides pointers on proper sample collection techniques and describes the most common errors that occur, especially in slurry sampling. It includes examples of ways to improve access for sampling in preparation plant design.

SAMPLING DEVICES

Proper sample collection starts with selection of the proper sampling device or tool and continues with knowing how to use the tool. The best way to sample and eliminate human error is to install mechanical samplers, the ultimate sampling tool. However, mechanical samplers at coal preparation facilities are a costly alternative to manual sampling and are generally reserved for payment samples. Considering how a mechanical sampler is designed and operated can reveal a lot about proper manual sampling. Manual sampling should try to mimic mechanical sampling.

Mimicking Mechanical Samplers

Mechanical samplers are designed to meet various national and international standards. Among other things, their design considers the top size of the material to be sampled, the best location for collecting a full stream of material or for taking a cut through the stream at a constant speed, and the size of the final sample that is needed. A variety of mechanical samplers are shown in Figures 1 and 2, representing both dry and wet samplers, respectively. Sample cutters are designed with a specific opening size to allow the largest particle to be collected. The sides are sharp. The cutters are parallel for a cross-stream sampler (used for belt discharges or launders) but are at an angle for samplers that rotate through a stream (used for nonpressurized pipes). This accounts for the difference in rotation speed between the outside and inside of the cutter. Another important feature is that the sample moves down a chute or pipe to a waiting collection device; it does not have a chance to bounce or splash out of the sampler.

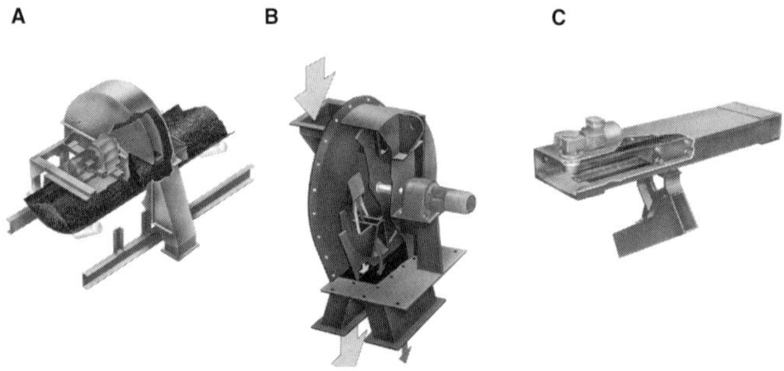

Courtesy of Multotec Process Equipment.

FIGURE 1 Types of samplers for dry bulk materials: (a) hammer sampler, (b) rotating plate divider, and (c) belt-end sampler

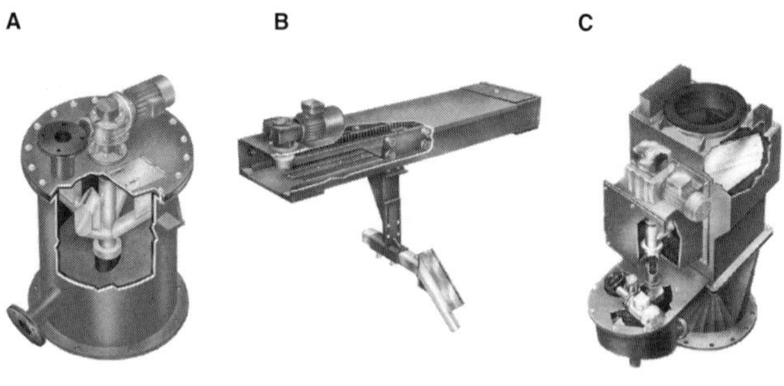

Courtesy of Multotec Process Equipment.

FIGURE 2 Types of samplers for slurries: (a) Vezin sampler, (b) launder sampler, and (c) two-in-one sampler

To mimic mechanical sampling, many coal sampling and analytical laboratories have developed specific tools of the trade. Several of these are shown in Figure 3. The tools feature sharp, parallel cutters with sufficiently large openings to allow proper sample collection. It is important to select a tool with a cutter that is at least 2.5 times the size of the largest particle. In addition, a tool should be selected that can contain the material sampled without allowing material to overflow.

Alternatively, a pipe thief can often be used for slurry lines with success, or a simple system can be set up that will allow the entire stream to be diverted into a sample container. This second system works only for streams that have low flow (up to 100 gpm [23 m^3/h]) and are nonpressurized. A sampling system using sample thieves for a column froth flotation circuit is shown in Figure 4. In this case, the flotation feed, froth concentrate, and froth tailing lines are equipped with thieves to collect a sample of slurry. The thieves are then connected with flexible hose to a sampling box. The sampling box is equipped

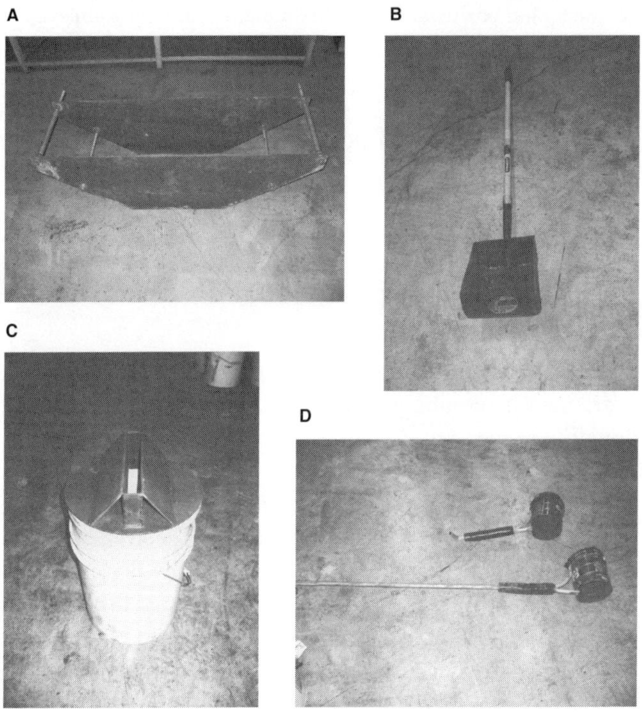

Courtesy of Standard Laboratories.

FIGURE 3 Tools of the trade: (a) stopped belt divider for 72-in. (183-cm), 35° belt, (b) flat shovel with built-up 4-in. (100-mm) sides, (c) slurry cutter for 5-gallon bucket, and (d) grab sample devices

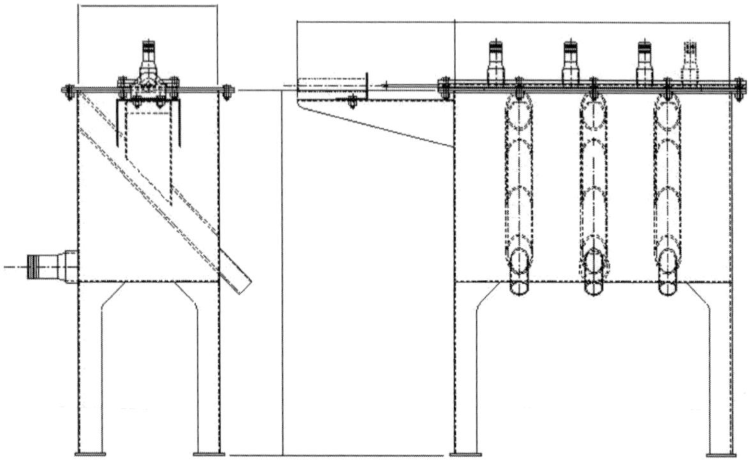

Courtesy of Clinch River.

FIGURE 4 Sampling box used with froth flotation column pipe thieves

with a mechanism that pushes the flexible hoses so that they discharge into sample buckets. In their parked position, the hoses exit the sampling box through a discharge pipe that is fed to the froth feed sump.

In some instances, shovels and buckets can be used effectively. However, they must be used properly, just as these more specialized sampling tools must be used properly.

Location and Planning

When a mechanical sampler is installed in a preparation plant, significant thought goes into determining its location. No less thought should go into determining the proper manual sampling location and preparing for sample collection. It is often difficult to find a proper location for sampling a given stream in a plant. The device chosen for sampling must be compatible with the sampling location and vice versa. Safety is also a major consideration. A review of the flowsheet can often lead to selection of alternative locations, and a walk through the plant with the flowsheet in hand is a must for determining the best locations. In addition, access doors may need to be installed, safety ladders may need to be used, and special sampling tools may need to be fabricated.

Sampling access can often be incorporated in chute design, as shown in Figures 5 and 6. Figure 5 shows an access door installed in a new discharge chute from a screen, and Figure 6 shows a port in a discharge chute from a screen bowl. Appropriate locations for sample doors should be considered in the initial design of a preparation plant.

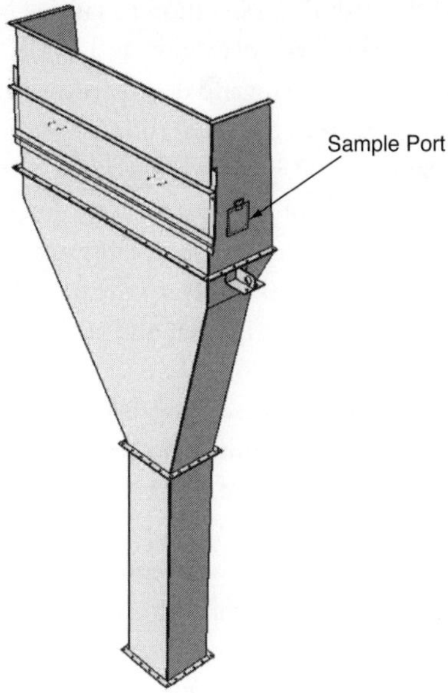

Courtesy of Farnham & Pfile Construction Company.

FIGURE 5 Sample door for double-deck screen discharge chute

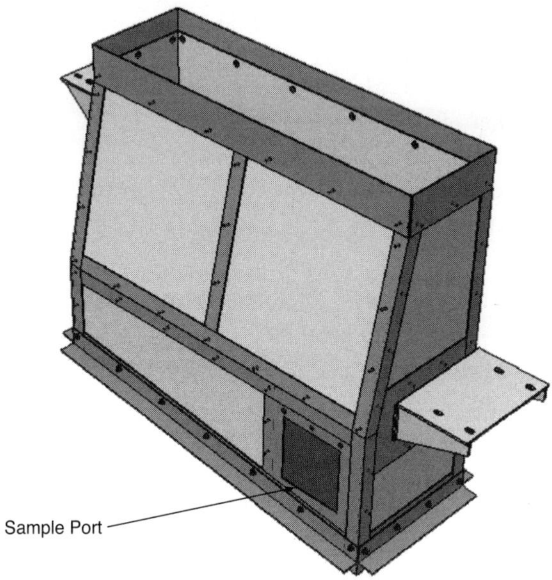

Courtesy of Farnham & Pfile Construction Company.

FIGURE 6 Sample door for screen bowl discharge chute

Perhaps the most difficult unit operation to sample is a bank of classifying cyclones. Where should the feed, overflow, and underflow streams be collected? Often the answer is upstream and downstream of the bank of cyclones. The flowsheet should be reviewed to determine whether there is a location where the entire stream can be sampled—perhaps at the desliming screen underflow for the feed sample or in the discharge lines of the overflow and underflow launders for the fine and coarse streams, respectively. The overflow and underflow may best be sampled as feed to the next downstream unit operation. Perhaps the overflow launder is open and each cyclone can be sampled. This may also be true for the underflow.

If the chosen sampling point is at the overflow from each cyclone, special care must be taken. A 20-in. (500-mm) diameter cyclone overflow can easily be more than 1,500 gpm (340 m^3/h) of slurry. Even a 14-in. (356-mm) diameter cyclone overflow can approach 1,000 gpm (227 m^3/h) of slurry. A quick pass of a bucket through this stream will easily fill it. It could also rip the bucket from the hands of the person collecting the sample.

The flow from a cyclone underflow is significantly less and is more manageable. However, cyclone underflow poses a different concern. For a properly operating cyclone, the underflow discharge is a spray that is often difficult to sample(see Figure 7). Care must be taken to collect a sample across this entire spray and to not preferentially collect any one portion of the spray.

To overcome difficulties in collecting cyclone feed samples, sample ports can be included in the initial distributor design and installed in the cyclone distributor, as shown in Figure 8. A small port with a valve allows a portion of the stream to be collected. Alternatively, the design can incorporate an extra opening with a shut-off valve (e.g., include a six-way distributor when only five cyclones are needed) that can be used to collect a full-stream sample. However, as mentioned previously, this may be a very high-volume stream, so care must be taken during collection.

USE OF MANUAL SAMPLING DEVICES

Proper use of the selected sampling tool is important. The cutter should be moved across the entire stream at a constant speed. If the sample overflows the container, it should be poured back into the stream and the process should be reattempted at a faster rate. The sides of the cutter should be placed perpendicular to the flow so that each particle has the same chance for entering the collection device. This is necessary whether the sampling tool is a shovel, bucket, or specially designed device.

A bank of conventional froth flotation cells poses a special problem. If the froth sample must be collected from the launder, it is important that the entire

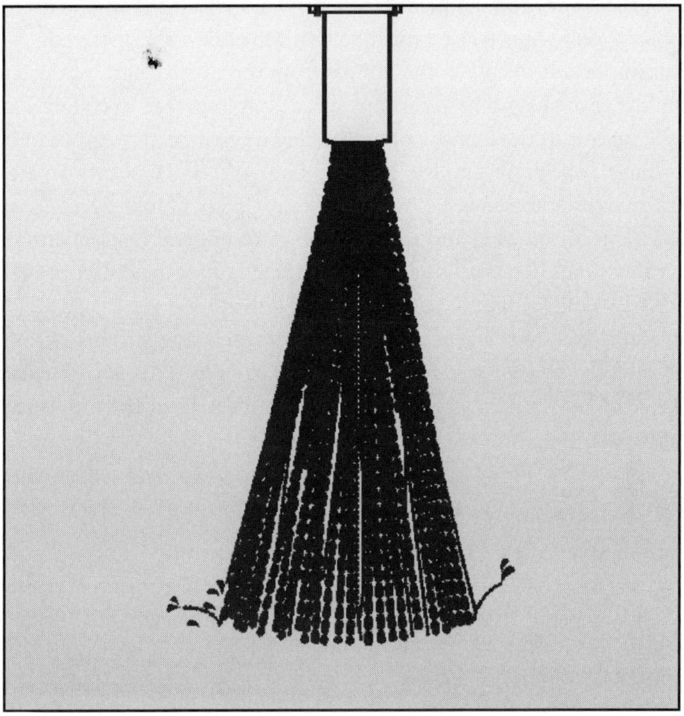

Courtesy of Multotec Process Equipment.

FIGURE 7 Underflow spray created by a properly operating cyclone

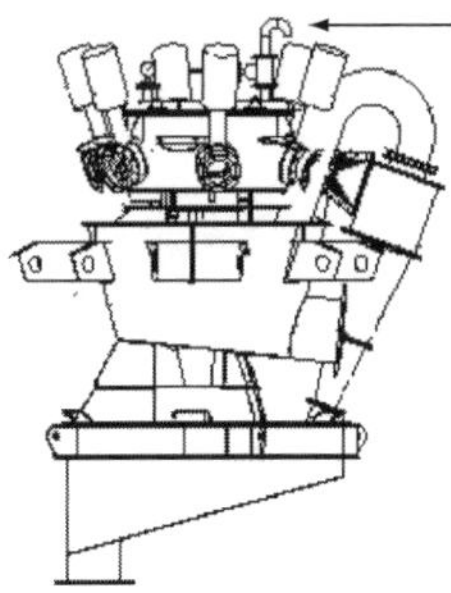

Courtesy of Multotec Process Equipment.

FIGURE 8 Classifying cyclone distributor with sample port

length of the launder be sampled. Froth is filled with a lot of air and often does not collapse easily. Special care must be taken to collect this material.

It is important to collect the sample from the entire length of the launder, walking the entire length at a constant speed. It is improper to collect a sample from the middle of the launder of each cell. The change in quality can be dramatic from the beginning to the end of a bank of cells, as shown in Figure 9. These data were collected at 1-ft intervals along a bank of four cells. The second cell had a dip in the weir, and excess water (and mineral matter) entered the froth at that point. If the cells had been sampled only at the center point of the launder, the results would have been skewed greatly.

In addition, never decant water or slurry from a sample. This introduces a bias toward the coarser fractions of the coal sample. This is true even with samples from streams that contain ultrafine slimes. Even these fine particles begin to settle.

SAMPLE CONTAINERS

Sample containers must be clearly and accurately identified. Improper labeling often occurs with sample containers when the bucket or drum is reused. For dums, a plastic protective envelope that adheres to the side of the drum and lid can be used. The contents should be clearly identified by writing the sample name, number, and date (and other pertinent information) on a piece of paper and placing the paper inside the protective envelope. This paper insert should be discarded when the sample is emptied from the container.

Plastic protective envelopes work well for buckets but can be a bit large for lids. A piece of duct tape can be applied to the side of the bucket and lid with pertinent sample information clearly written on the tape with a permanent marker. When the sample container is emptied, the tape can be removed and the surface is then free of extraneous information.

CONCLUSIONS

Sample collection is an important part of any coal preparation plant operation. The data generated are only as good as the sample collection method.

For proper sample collection,

- Try to mimic mechanical sampler operation:
  - Take a full stream cut.
  - Use a constant speed.
  - Use parallel cutter blades for cross-stream sampling and tapered cutters for sampling pipes (radial).

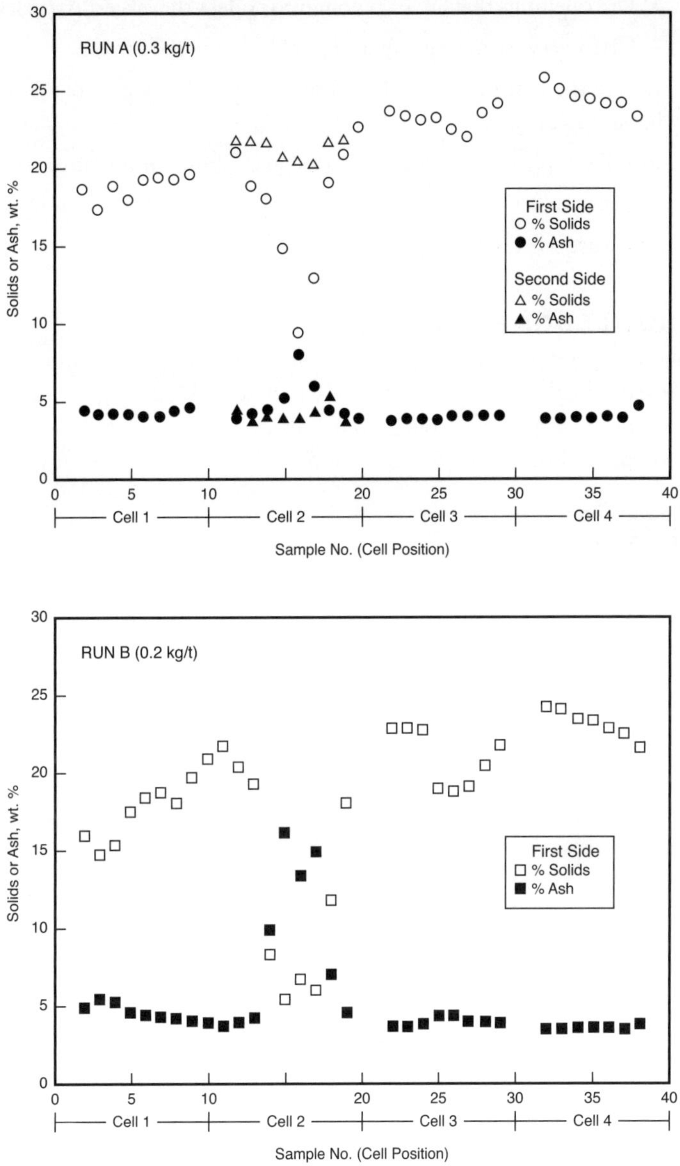

Source: Arnold and Aplan 1990.

FIGURE 9 Percentages of solids and ash from samples collected at 1-ft (0.3-m) intervals along a bank of conventional froth flotation cells

- Use openings that are big enough to collect the largest particles.
- Give every particle an equal chance of being collected.

- Locate the best place to collect a sample, provide adequate access, and insist on safety.
- Install sample ports in the initial design of chutes or distributors.
- Use the proper sampling tool.
- Take care when labeling samples.

ACKNOWLEDGMENTS

The authors gratefully acknowledge the contributions of pictures and drawings from Standard Laboratories, Inc., South Charleston, West Virginia; Multotec Process Equipment, Johannesburg, South Africa; Clinch River, West Virginia; and Farnham & Pfile Construction Company, Inc., Monessen, Pennsylvania.

REFERENCE

Arnold, B.J., and Aplan, F.F. 1990. Effect of water carryover in a bank of industrial coal froth flotation cells with a comparison to laboratory test results. *Min. Eng.* 42(12):1347–1351.

Coal Analyzers Applied to Coal Cleaning: The Past, Present, and Future

Richard Woodward

ABSTRACT

Coal preparation plants are not renowned for their use of sophisticated instrumentation, so it is surprising to find them using on-line coal analyzers, perhaps the most expensive instrument known in the coal industry. This can be a good decision in some cases and a waste of money in others. This chapter explains when on-line analyzers can benefit the user and describes the advances made in on-line coal analyzers in the quarter century since their introduction. Different types of coal analyzers are explained, and guidance on when to use which type is provided.

INTRODUCTION

Coal preparation plants first began using on-line coal analyzers in the early 1980s, usually to monitor the ash content of product streams and keep ash levels close to their target values. Australia and the United States led the way in adopting this means of control, but by the mid- to late 1990s usage had become global, with China, India, and South Africa using large numbers of analyzers. The device manufacturers initially were from Australia, the United States, and Germany, but eventually there were local producers in India, China, and South Africa. The use of analyzers for plant control has declined in the current decade, primarily because fewer plants are being built today, and there has been market saturation among existing plants.

Most analyzers used in coal preparation plants are simple ash gauges, not surprising given the fundamental objective of coal cleaning, which is to remove ash. A smattering of elemental analyzers are used, but their cost is much higher, making a quick payback more elusive. Most analyzers are located on the "clean side" of the plant, but more than 100 are also found on the plant feed. Almost all plants use a "man in the loop" rather than automation to interpret the analyzer results and effect a process change. Furthermore, the most common means of controlling ash in response to changes in the product's ash content is to adjust heavy-media gravities. This chapter explores these methods in greater detail and offers recommendations to future buyers of analyzers.

ANALYZER TYPES

Two primary types of on-line coal analyzers are found in coal preparation plants: ash gauges and full elemental analyzers. As their name implies, ash gauges measure only ash. Two common techniques are dual gamma (also called low-energy transmission or dual-energy transmission) and natural gamma. Dual-gamma ash gauges use two gamma-ray sources (usually americium 241 and either cesium 137 or barium 133) of differing energy levels and measure the relative attenuation of the gamma rays passing through the coal. The attenuation of the high-energy gamma ray is roughly proportional to the mass of material between the source and the detector, and the lower-energy gamma ray is attenuated differently according to the atomic number of the atoms between the source and detector. The lower-energy gamma ray is attenuated more by the elements found in coal ash and not as much by carbon, hydrogen, nitrogen, and oxygen. When the two gamma sources are used in combination, one tends to act as a normalizer, accounting for mass flow changes, whereas the other can detect variations in ash level.

The other less common ash gauge is the natural gamma gauge. This analyzer relies on the tendency of some coals to possess, as a fairly constant part of the ash, isotopes of thorium or potassium, which are natural gamma emitters. Thus, as the ash levels increase, the amounts of these gamma rays and the amount of their signal received by the detector also increase.

Technical limitations exist on the accuracy of each of these ash gauges. The dual-gamma ash gauge is very sensitive to, and incapable of detecting, changes in the percentage of ash that is iron oxide (Fe_2O_3). A change in the Fe_2O_3 percentage of ash as little as 1% in a 10% ash bituminous coal will cause an error in total ash determination of about 0.6%. This phenomenon is proportional to the total ash amount, so for a typical raw coal of 25% ash, the impact is far greater—about 1.5% (Figure 1).

For this reason it is difficult to understand the utility of ash gauges being applied to the feed stream for a coal preparation plant, as will be discussed in more detail later.

The obvious limitation of the natural gamma ash gauge is it relies on the assumption that the naturally radioactive potassium or thorium is a constant percentage of the ash. This assumption is clearly untrue, but how much it varies is unknown.

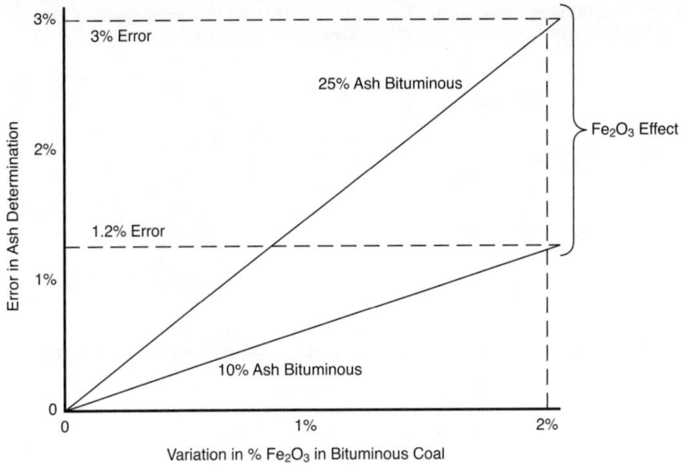

FIGURE 1 Ash measurement error in dual-gamma ash gauges caused by variations in Fe_2O_3 fraction of ash

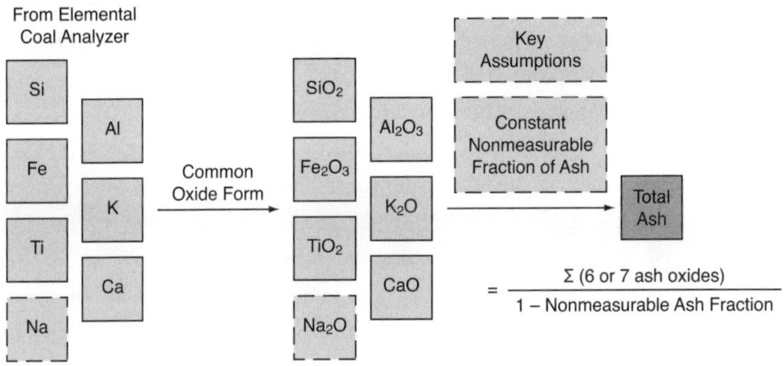

FIGURE 2 The process by which elemental analyzers measure ash

Most elemental analyzers use a technology known as prompt gamma neutron activation analysis (PGNAA) to measure up to 11 elements in coal, including the major ash constituents: silicon dioxide (SiO_2), aluminum oxide (Al_2O_3), Fe_2O_3, calcium oxide (CaO), titanium dioxide (TiO_2), potassium oxide (K_2O), and, in some cases, sodium oxide (Na_2O) (Figure 2). In bituminous coals, the first six ash constituents typically account for 95%–96% of the ash, so by measuring the concentrations of these six constituents and scaling up to 100% by a factor characteristic of the mine, total ash is determined. These PGNAA analyzers use neutrons to activate the coal, and the gamma

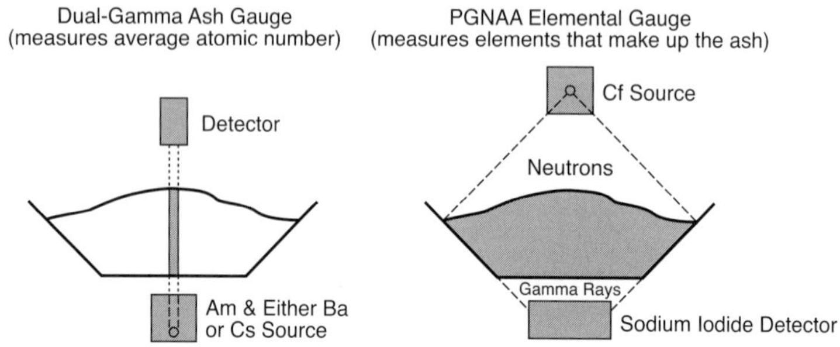

FIGURE 3 Comparison of dual-gamma ash gauge and PGNAA analyzer scans

rays emitted have energy levels that differ according to the element involved. By detecting the gamma rays and constructing a composite spectrum of energies, it is possible to determine the elemental composition of the coal.

PGNAA analyzers are more accurate than ash gauges because they are not vulnerable to changes in the Fe_2O_3 fraction of the ash, and they analyze the entire cross-section of coal rather than a thin sliver (Figure 3).

Elemental analyzers are also more expensive, sometimes approaching $500,000. There are two types of these analyzers: one that mounts around the conveyor (Figure 4) and another that analyzes sample streams (Figure 5).

PGNAA units that mount around the conveyor are less expensive than sample stream units, but they are not as accurate. Sample stream analyzers benefit from having an optimal, constant-geometry cross-section to analyze.

The most commonly found analyzer at coal preparation plants is the dual-gamma ash gauge. An analysis of a year 2000 reference list of Scantech, the leading supplier of this instrument, and a current reference list of Thermo Scientific shows that slightly more than half of Scantech's ash gauges have been sold into coal preparation plant applications, as have about a third of their elemental analyzers. In contrast, Thermo Scientific's elemental analyzers are found in coal preparation plants only 15%–20% of the time (Figure 6). The majority of Thermo Scientific's coal analyzers are used for blending at the coal mine or power plant, along with boiler optimization at the power plant.

METHODS OF PROCESS CONTROL

As mentioned earlier, there are different ways to respond to the minute-by-minute coal quality information received from the on-line analyzer. If the operator believes that the analysis is reliable and the observed trend is not just a random event, the most common response is to increase the heavy-media gravity to increase ash content or decrease it to reduce ash levels. This has to

FIGURE 4 **Thermo Scientific elemental crossbelt full-flow analyzer at Sierra Pacific's North Valmy power plant**

FIGURE 5 **Thermo Scientific Coal Quality Manager sample stream analyzer at the Alliance Coal Gibson County coal preparation plant**

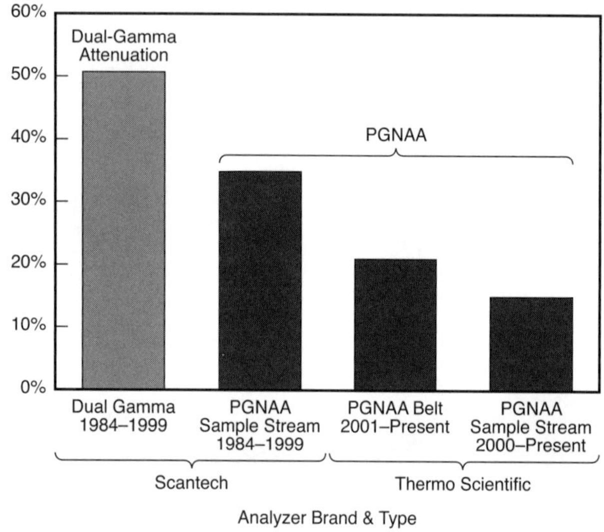

FIGURE 6 Preparation plant applications as a percentage of all analyzers sold

be viewed as a coarse control strategy because the time constant for new gravity settings to be reached is on the order of 15 minutes. If this is the current method used, gravity settings should not be varied any more often than hourly, preferably only once or twice per shift. In addition to the time constant to effect a gravity change, complications arise from the fact that the analyzer itself can produce random error. If the analyzer is a dual-gamma ash gauge, a 1-hour one-sigma error of 0.5%–1.0% is not unusual. If the indicated ash value has risen by 1%, such increase may be significant enough to warrant a gravity change. Furthermore, random variations of 2% or higher on minute-by-minute results are not unusual, and the operator must be careful not to respond to those events.

A second, more effective method of controlling product ash is to use a continuously variable flop gate on the feed to the plant. This flop gate diverts a fraction of the raw coal around the plant to maintain a constant ash level in the product (Figure 7).

For example, if the target ash level is 10%, raw ash is 25%, and the ash level in the clean coal coming out of the plant is 8%, a bypass fraction of 11.8% will achieve the target. If the raw ash level increases, the bypass fraction can be reduced, and vice versa. This method is easily understood and does not involve any changes in heavy-media gravity. It is regrettable that few plants have moved in this direction, although one that did is discussed elsewhere (Goshert et al. 2006).

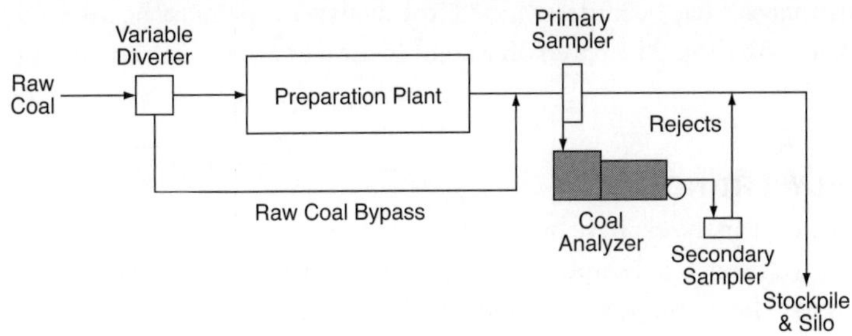

FIGURE 7 Coal flow schematic for variable raw coal bypass to control ash quality

A third, seldom seen method of control can be found in some modern jigs. If it is possible to adjust the refuse gate height, the plant can respond to variations in the product ash in that manner.

INSTALLATION LOCATION

In this chapter's introduction, it mentioned that several coal preparation plants, though fortunately not a majority, have ash gauges on both the plant feed and product streams. Perhaps the low price of ash gauges (at least compared with elemental analyzers) coupled with a zeal for absolute control over quality led to these decisions, but it is unlikely that a return on investment in on-line analysis of the plant feed will ever be realized. An ash gauge can be inaccurate on high-ash coals, and predicting the quality of the product based on the quality of the feed is an inexact science. There are days when a 25% raw coal with a specific gravity of 1.50 will produce an 8.5% product ash, and other days that figure will be 10%. Putting an on-line analyzer on the feed to the plant is not recommended.

Whether a plant needs an analyzer hinges largely on what happens to the coal downstream of the plant. When the product quality is on the desired side of the blended quality set point, it may not be necessary to have an on-line analyzer at the plant. For instance, if all the clean coal is conveyed to a clean coal silo where it will be blended with a coal or coals of differing and inevitably lower quality, then an analyzer might not be necessary.

The other extreme would be that the product goes directly to some storage site (silo or stockpile) from which its reclaim is not blended. In essence, the product quality emanating from the plant is exactly what a customer will receive and will be used to determine whether the coal producer has met the contract specification. In such cases it would be shocking for the coal producer not to install an analyzer on the product stream of the plant. Once the need for an analyzer has been determined, the analyzer type must be selected, and that selection should be based on a combination of accuracy needs and budget constraints.

LIKELY TRENDS

On-line coal analyzers have been in use for almost a quarter century. Although there have been steady improvements, none can really be considered breakthroughs. These improvements include the following:

- Digital electronics
- More robust calibration methods
- More extensive and functional displays on the operator console
- Better methods of determining instrument accuracy
- Smaller footprints
- Automatic diagnostics
- Lower source strength needs

The most significant advancement in on-line coal analysis has probably been the introduction of elemental crossbelt analyzers early in this decade (Empey et al. 2003). Crossbelt analyzers have carved out a significant niche, and at this point they outsell sample stream analyzers.

Where are we likely to see advances in the next quarter century? Some have predicted major advances such as smaller, better-performing units with less source radiation. Continuous improvement is certain, but there are physical limits to how small an analyzer can be without compromising performance or safety. Most of an analyzer's size comes from shielding for the safety of operators and service engineers. If that shielding were to be further reduced, it could only occur with decreases in source loading (which are possible, but there is potential for performance degradation) or by increasing levels of radiation.

Advances will probably come in other areas. Autodiagnostics will mean greater uptime, allowing operators to detect problems sooner and make repairs before performance or availability is compromised. Electronics will continue to get smaller, allowing the use of smaller cabinets. As analyzer suppliers stay attuned to advances in detectors, neutron generators, and spectral processing methods, cost-effective improvements will be made wherever possible. But the reality today is that on-line coal analyzers are underused and incremental improvement is not likely to cause new buying trends. The greatest advancement that could occur with on-line analyzers is for coal producers to realize that these instruments are available today with excellent accuracy and reliability at prices that offer attractive returns on investment to the smart user.

SUMMARY

Many coal preparation operators have used on-line coal analyzers during the past 25 years to improve control over product quality. A variety of analyzer types, locations, and control strategies are available. This chapter has shown when an analyzer at the coal preparation plant makes sense, what types are preferred, where they should be located, and how their on-line analysis information should be used. When used wisely, analyzers can be a valuable addition to a plant.

REFERENCES

Empey, E., Evans, M., and Woodward, R. 2003. A major step forward for on-line coal analysis. In *Proceedings of Coal Prep 2003*, Lexington, KY. Stamford, CT: Primedia Business Exhibitions. pp. 137–151.

Goshert, M., McKenzie, L., and Woodward, R. 2006. Quality control made simple. *World Coal* (May):19–22.

Useful Instruments Developed for CSIRO Coal Preparation Projects

M. O'Brien and B. Firth

ABSTRACT

In research projects performed at the Commonwealth Scientific and Industrial Research Organisation in association with the Australian Coal Association Research Program, several instruments were developed to accurately measure important variables such as screen motion, panel aperture, and flow rates in normally difficult-to-measure flow streams. This chapter discusses the use of these instruments in coal preparation and their subsequent commercialization.

BACKGROUND

Researchers at Australia's Commonwealth Scientific and Industrial Research Organisation (CSIRO) in Brisbane have developed several instruments aimed at easily measuring variables such as screen aperture, screen motion, and flow rates in lined pipes. These instruments were developed during research projects in which a need arose to accurately measure a variable that was not normally monitored easily. These instruments were effective in providing the information needed for these projects, but they also were essentially fulfilling the proof of concept for the development of potential new commercial monitoring devices. This chapter describes the three new developments and their capabilities.

APERTURE AND OPEN AREA

Australian Coal Association Research Program (ACARP) Projects C7048 (Myers et al. 2000) and C8042 (O'Brien et al. 2002) highlighted the need for an improved method of measuring screen aperture and open area inclusive of pegging to measure screen wear characteristics. In both projects screen efficiencies were highly correlated to the screen aperture (and thus wear) and open area (Figure 1). Few plants currently monitor the wear of their screens in terms of aperture size and open area, factors that have a major influence on magnetite losses on drain-and-rinse (D&R) screens and yield losses due to misplaced material on desliming screens. In addition, plants rarely check new screens to see whether they are within specifications.

ACARP project C11006 (O'Brien et al. 2003b) led to the development of a portable prototype measurement system using a digital camera and a laptop computer to measure screen aperture and open area. The measurement of the screen aperture included the number of aperture measurements made, the mean, the histogram, and the standard deviation of those measurements. The system was also capable of measuring the open area inclusive of pegging.

Hardware and Software

Pixel resolution and number of aperture slots measured (size of measurement area) were the main drivers for the selection of the camera resolution and the design of the camera mount. Figure 2 shows the prototype camera mount and image mask. The camera selected had a 5-megapixel resolution, which translated to approximately 14 pixels in width per 0.5-mm aperture, covering an area on the screen deck 10 cm × 7 cm at a focal length of 0.27 m. Approximately 20–30 aperture slots (depending on aperture size and duty) were measured using this method, with 60 measurements carried out along the lengths of each slot. Two

FIGURE 1 Measurement of screen apertures using an optical method

external flash units were used to provide the illumination, and LabView image analysis software was used to analyze the images. Figure 3 shows the software output from the system showing the histogram and measurement parameters, which are saved to a file. The raw measurements (up to 1,000 points) are also saved for later data analysis if needed.

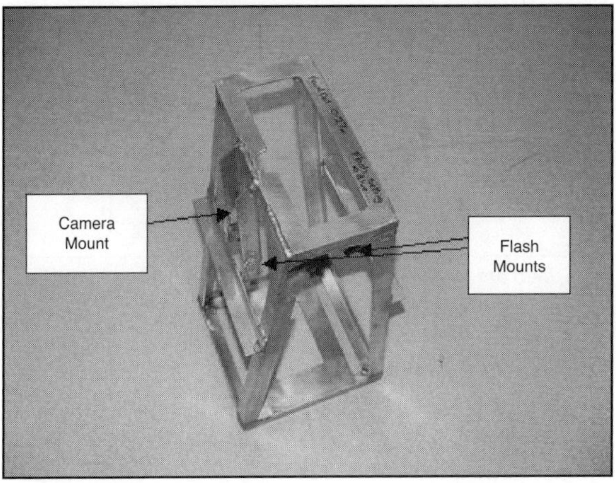

FIGURE 2 Camera and flash mounting support structure incorporating an image mask and calibration points

FIGURE 3 Software output showing histogram and measurement parameters

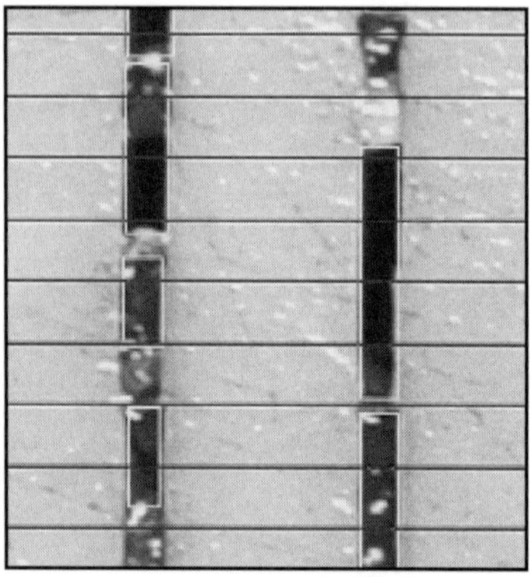

FIGURE 4 Cropped area of an ROI showing open apertures identified as particles

Open area measurements were carried out on the region of interest (ROI) based on the size and number of slots. Pegging measurements were determined using a particle analysis technique that identified the clear slots as particles, as shown in Figure 4.

Method Validation

Validation of the method was performed on two new panels of apertures 0.5 mm and 0.8 mm and a used 0.5-mm panel, with optical, taper gauge, and image analysis techniques used for comparison. During a previous Australian Mineral Industry Research Association project (Firth et al. 1995), the Grubbs estimation technique (Grubbs 1948, 1982) was used to determine the bias and precision of a measurement method. This technique is the basis of the current Australian Standard (AS 1038.24-1998) for determining the precision and bias of on-line instruments in the coal preparation industry. The results of the Grubbs estimation technique applied to the three measurement methods are shown in Table 1. In all cases the use of image analysis was found to be the most precise.

Plant Testing

On completion of the validation at Queensland Centre for Advanced Technologies (QCAT), the system was tested at a coal preparation plant on a nominally 0.5-mm product and reject D&R screens and desliming banana screens after normal optical measurements by the screen maintenance contractor. This work was completed with the assistance of consultants A&B Mylec, who then independently evaluated the system at two other Bowen Basin plants on low-head screens.

TABLE 1 Results of the Grubbs estimation for the three techniques

	Screen 1, 0.5 mm (New)			Screen 2, 0.8 mm (New)			Screen 3, 0.5 mm (Used)		
	Taper Gauge, mm	Portable Optical Micro-scope, mm	Image analysis, mm	Taper Gauge, mm	Portable Optical Micro-scope, mm	Image analysis, mm	Taper Gauge, mm	Portable Optical Micro-scope, mm	Image analysis, mm
Variance	0.001	0.001	0.001	0.003	0.002	0.001	0.019	0.016	0.020
Precision	0.053	0.050	0.038	0.088	0.057	0.029	0.174	0.214	0.184
Mean	0.55	0.56	0.55	0.97	0.87	0.81	0.89	0.91	0.84
Standard deviation	0.02	0.03	0.02	0.04	0.03	0.02	0.10	0.11	0.08

Three sites in the Bowen Basin were selected for testing of the digital imaging technique. The sites were selected to allow testing across a broad range of fine aperture screening applications and screen deck designs. All site testing was completed on screening applications with a nominal aperture of 0.5 mm.

The site testing was coordinated with the regular maintenance period of the coal processing plants. Following plant shutdown, the screen decks were cleared of coal, as would be done for regular maintenance work. On most occasions, additional cleaning of the decks was required to clear access to the screen deck itself and to avoid inaccuracies in the image analysis. Because the technique measures the open area relative to the total area of the analyzed image, the presence of free coal particles on the deck surface in the digital camera field of view introduces errors into the open area determination.

After screen deck preparation, digital images of each screen panel were collected. Initially, up to three images per panel were taken; however, this was reduced to one per panel because of time constraints.

Field Trial Results

The digital imaging technique can be used to measure the screen aperture and open area of the screen panel section captured by the image. To extrapolate the results to reflect the properties of the total screen, corrections must be made for screen regions not captured by the images. Such regions include structural support in each panel and blanked areas in the feed and discharge zones and around the hold-down mechanisms.

Essentially the total screen panel area can be differentiated into a screening component and structural support component (Figure 5). In most cases the dimensions of the overall screening and support areas can be measured conventionally and are uniform for panels of the same design.

The digital imaging technique is also used to measure the effective open area of the screening area by selecting a uniform region within each image for analysis (Figure 6). If the eventual region analyzed is a representative sample of

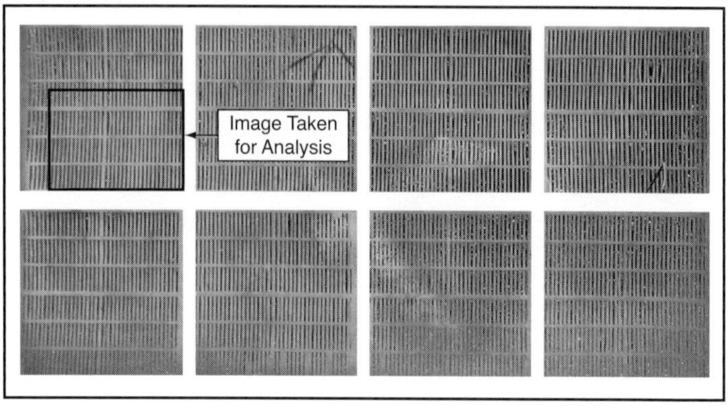

FIGURE 5 Screening and structural support areas of screen panel showing region captured for open area and aperture determination

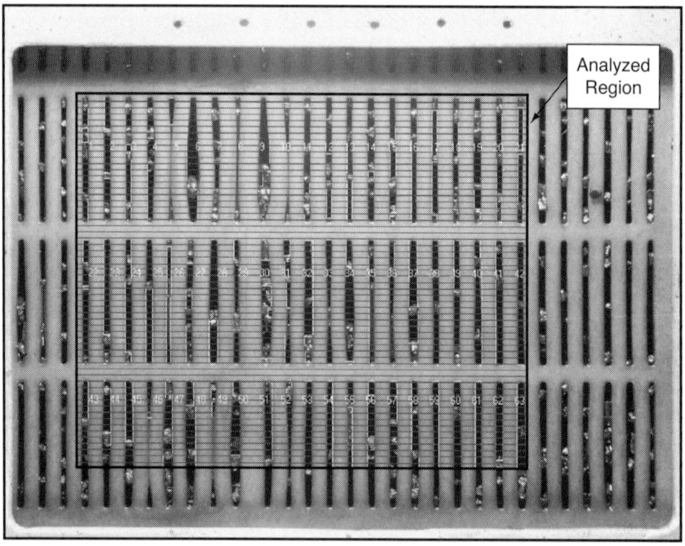

FIGURE 6 Analyzed region as part of total screen image

the total screening area of the panel, the open area for each screen panel can be calculated. Estimations of open area and screen aperture would be needed for panels that proved to be inaccessible to the testing apparatus.

The total open area for the entire screen, therefore, is the sum of the calculated open area for the individual screen panels. Similarly, the average aperture opening is the weighted average aperture from each screen panel based on the effective open area. The formulas for open area and average aperture determination are as follows:

$$\text{actual open area} = \sum_{i=1}^{n} \text{OA}_i$$

$$\text{OA}_i = \%\text{OA}_i \times \text{SA}_i$$

$$\text{average aperture} = \frac{\sum_{i=1}^{n} (\text{OA}_i \times \text{Ap}_i)}{\text{actual open area}} \qquad \text{(EQ 1)}$$

$$\text{effective open area (\%)} = \frac{\text{actual open area}}{\text{total screen area}}$$

where

actual open area = total screen actual open area (m^2)
n = number of panes in total screen
OA_i = open area for ith screen panel (m^2)
$\%\text{OA}_i$ = percentage of open area result for ith screen from digital imaging technique
SA_i = screening area, as measured, for ith screen panel (m^2)
average aperture = average screen aperture (mm)
Ap_i = average panel aperture for the ith screen panel (mm)
effective open area (%) = calculated percentage of open area for entire screen
total screen area = total screen area as measured (m^2)

From the evaluation, the full screen results were determined and are shown in Table 2.

In addition to providing total screen results, the panel-by-panel measurements allow useful analysis of changes in screen panel condition within the screen. Figures 7 and 8 are provided as examples of how a profile of a screen can be generated for the important screen characterization measurements through appropriate data smoothing techniques. Figure 7 shows the screen aperture profile for Screen 3 at Site 1, highlighting how the screen apertures are larger at the feed end of the screen, presumably as a result of greater wear. Similarly, the pegging profile of Screen 3 at Site 2, shown in Figure 8, highlights how the degree of pegging increases along the length of screen. As these

TABLE 2 Calculated results using the described method

	Application	Average Aperture, mm	Actual Open Area, m^2	Total Screen Area, m^2	Effective Open Area, %
Site 1					
Screen 1	Desliming	0.72	2.00	29.8	6.7
Screen 2	Product D&R*	0.72	1.52	29.8	5.1
Screen 3	Product D&R	0.73	1.44	29.8	4.8
Screen 4	Reject D&R	0.72	1.13	15.9	7.1
Site 2					
Screen 1	Product D&R	0.75	1.28	14.8	8.6
Screen 2	Dewatering	0.67	0.54	8.9	6.2
Screen 3	Reject D&R	0.64	0.62	8.9	6.9
Screen 4	Desliming	0.61	1.86	14.8	12.5
Site 3					
Screen 1	Product D&R	0.68	0.99	11.9	8.3
Screen 2	Product D&R	0.67	0.99	11.9	8.3
Screen 3	Reject D&R	0.67	1.03	11.9	8.7
Screen 4	Product D&R	0.70	1.46	11.9	12.3

*D&R = drain and rinse.

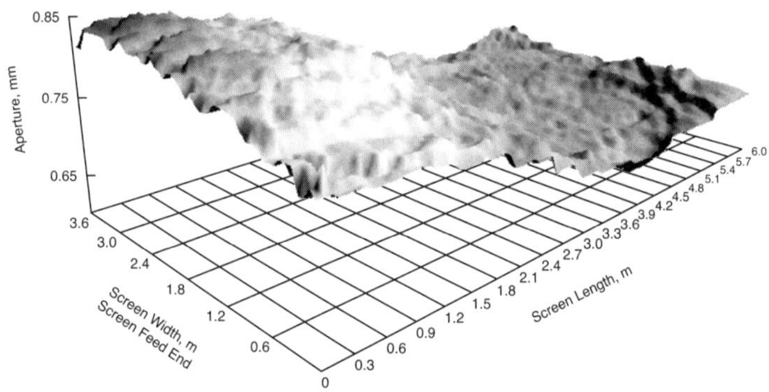

FIGURE 7 Screen aperture measurements for Site 1, Screen 3

examples show, the application of the digital imaging technique to provide comprehensive measurements of screen decks can allow new approaches to screen deck characterization.

Histograms

The histogram of the aperture measurements is a convenient visual method to describe the variation in measurements. The histogram can be produced for individual panels, for the entire screen, or for screen sections according to the

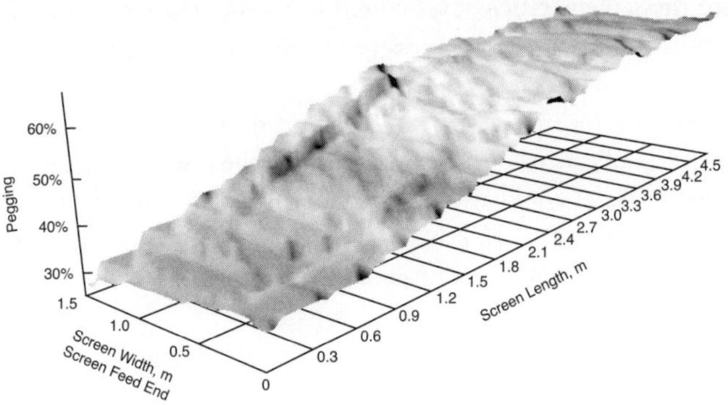

FIGURE 8 Screen pegging measurements for Site 2, Screen 3

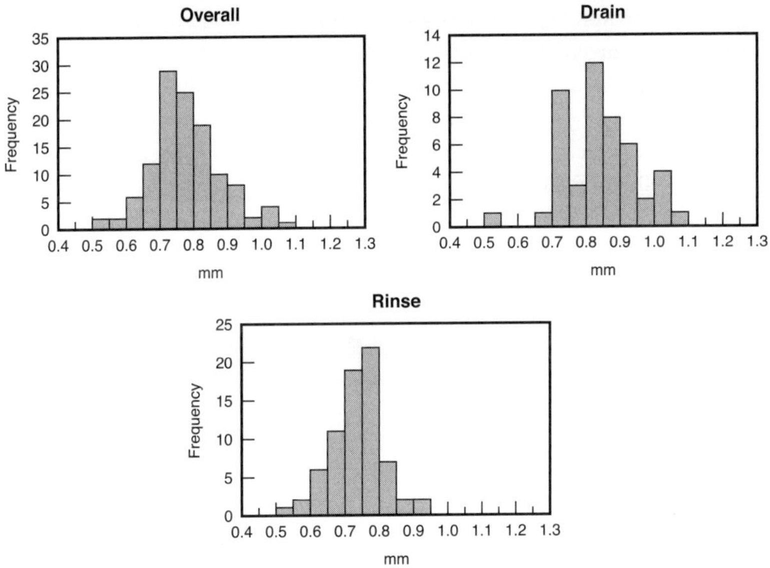

FIGURE 9 Aperture histograms produced from Site 1 product screen

standard deviation and can be used to visually observe panel wear. Figure 9 shows histograms produced from the measured apertures of a product's D&R screen at Site 1. These histograms could be used to monitor screen wear patterns over the life of the screen panels (histogram with time) and, when related to the screen's efficiency, be used to predict the screen panels' critical failure point (point at which excessive misplaced material has a major effect on screen and plant efficiency).

For instance, individual panel histograms showing a non-Gaussian distribution would indicate uneven wear. Similarly, if the histogram showed the presence of two peaks, it could also indicate uneven wear or teardrop or localized wear on an area of the panel. It is also conceivable that a histogram with more than one peak may indicate poor construction materials, with some areas of the screen wearing preferentially.

Wear and Tonnage

Two panels—one each from a high-wear and a low-wear area—were chosen from a screen at a Bowen Basin plant to demonstrate how individual panels performed with time and tonnage. The screens are processing a hard coking coal with a Hardgrove grindability index of approximately 85, at 400 t/h, with a top size of 16 mm. Figure 10 shows performance results for a low-wear panel that was not changed during the trial period, and Figure 11 shows results for a panel that was changed twice during the trial. These results show that at this plant, panels in high-wear areas are changed after approximately 200,000 tonnes of screen feed, and panels in low-wear areas are changed after approximately 400,000 tonnes of screen feed.

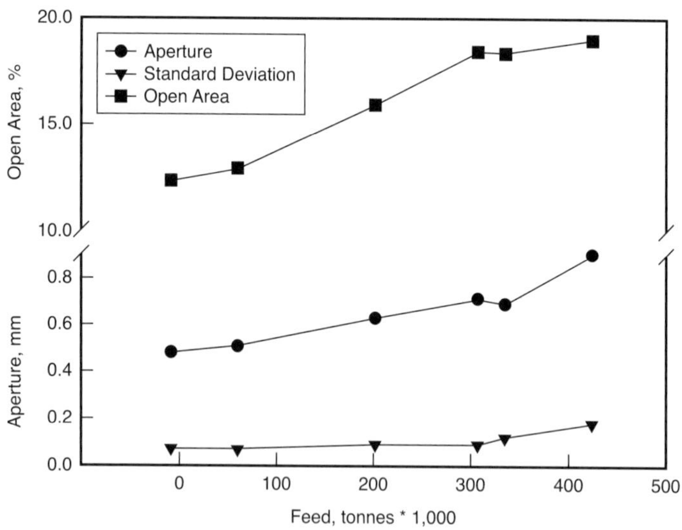

FIGURE 10 Panel C3 from Screen 4 showing a low-wear area

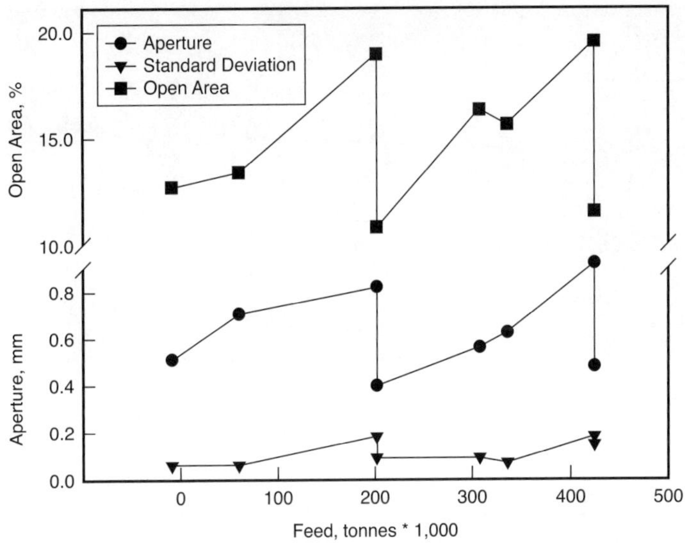

FIGURE 11 Panel D5 from Screen 4 showing a high-wear area with panel changed twice during trial period

PORTABLE SCREEN MOTION ANALYZER

CSIRO developed a portable PC-based system to meet the needs of ACARP project C8042, which incorporated triaxial accelerometers and produced a display of the motion of a banana screen in three dimensions together with frequency and stroke. This system was hardwired to accelerometers mounted on the screen with magnets and was not robust or portable enough for permanent use as an instrument. In a second ACARP-supported project (O'Brien et al. 2003a), the objective was to produce a prototype robust, portable, handheld measurement system capable of measuring, displaying, and recording the screen motion in terms of frequency, stroke, and three-dimensional (3-D) movement.

A brief literature search showed that many vibration instruments were available; however, the majority were designed to measure bearing vibration or shaft movement for maintenance purposes only. Only two manufacturers—SKF (U.K.), with their Copperhead range, and Brüel & Kjær Schenck's condition monitoring systems—offered instruments that were specifically designed as screen fault detection systems. These systems concentrate on bearing-condition monitoring but can also measure the overall screen vibration and can be equipped with an alarm for cutoff when the vibration exceeds set ranges. Although they measure vibration, these devices do not provide output for frequency or stroke and do not attempt to quantify the screen motion.

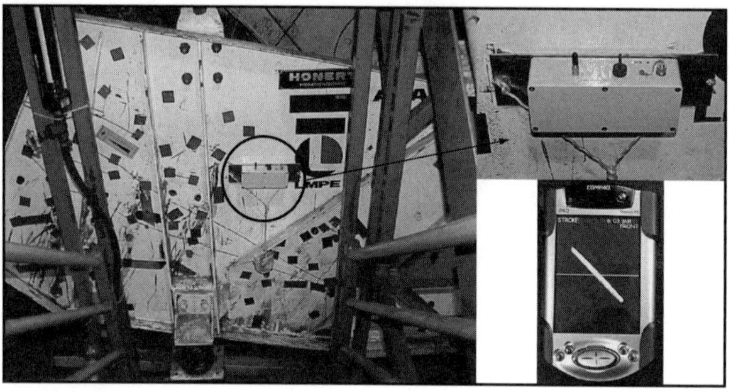

FIGURE 12 Accelerometer package attached to pilot banana screen and output on pocket PC

Hardware

The movement of the screen is measured with a 3-D accelerometer system. A ConnectBlue serial port adapter is configured to connect to any Bluetooth peer, and a Compaq pocket PC is used to connect to the transducer on the screen. All switches and connectors are waterproof, and the enclosure is attached to the banana screen with rare earth magnets.

A simple communication protocol between the pocket PC and the Atmel Mega128 microprocessor enables up to 10,000 data points at 1,000 samples per second for each of the three accelerometer axes to be sent. An image of the accelerometer package attached to the pilot banana screen at CSIRO is shown in Figure 12.

Software

The software on the pocket PC performs a double integration on the accelerometer data to give position versus time in three dimensions. The stroke is calculated from the positional data of the x- and y-axis motion, and the frequency is determined by averaging the time between peaks. The results, including a 3-D vibration graph, are then displayed. The data sets can be stored for later downloading to a PC. The software for the pocket PC was written in embedded C^{++} and later in LabView.

Stroke Experiments

Stroke is the measure of distance traveled in the direction of screen excitation and is calculated from the measured displacements in the x and y directions. Stroke was measured using the instrument 44 times at a medium stroke (6 mm)

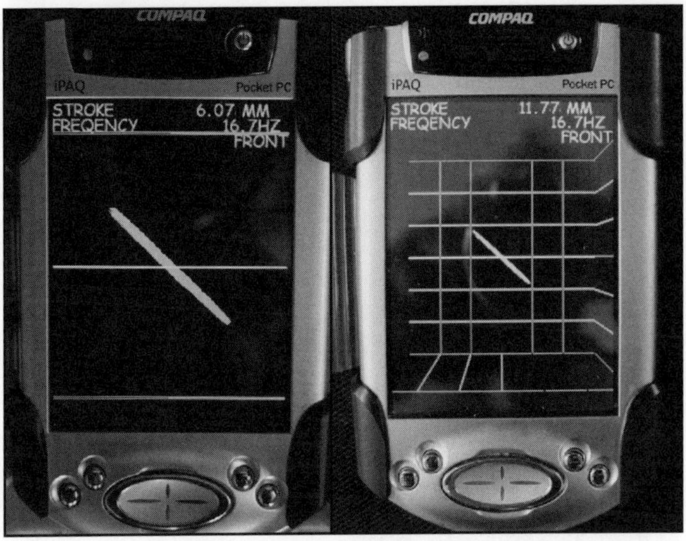

FIGURE 13 Screen shots of two stroke experiments, with smaller stroke enlarged (space between grid lines is 5 mm)

and maximum stroke (14 mm) to determine the standard deviation of the measurements. The standard deviation was 0.12 mm on a mean stroke measurement of 6.08 mm and 0.021 mm on a mean stroke measurement of 11.77 mm (approximately 0.2% relative). Figure 13 shows the screen outputs for strokes of 6.08 mm and 11.77 mm, respectively.

Frequency and Motion Experiments

The frequency of screen vibration on the pilot screen was adjusted by using the variable speed drive so that the output and motion could be observed using a portable motion analyzer as the driven frequency was changed. Figure 14 shows the screen outputs from the linear experiments and Figure 15 from a normal operation. Figure 14 shows a series of screen shots taken from the analyzer, with the initial start shown in the top left-hand corner followed by normal motion at 16 Hz to the left and progressing to 6.3 Hz to the bottom right. As the driven frequency was reduced, the motion tended to move away from the normal linear pattern toward a rotary motion, as shown in Figure 15.

Figure 16 shows the effect of placing a weight on the rear tab of the screen. This had the effect of offsetting the motion in both the y and z directions.

Sixteen screens at a Bowen Basin plant were examined using portable screen motion analysis to observe the loaded operating motion of the screens.

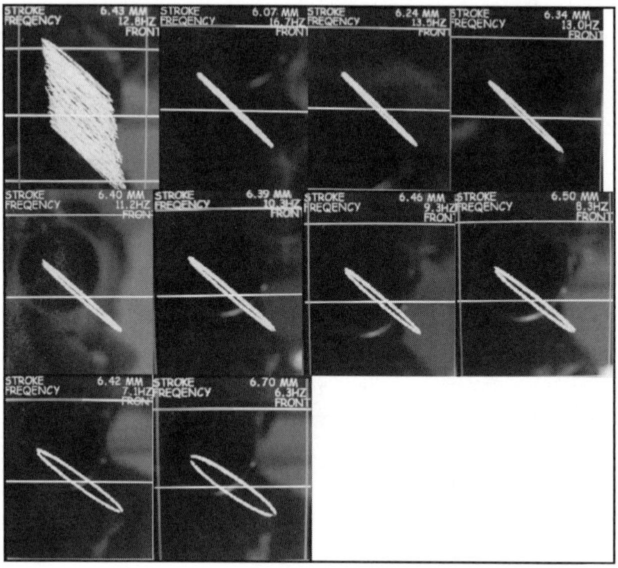

FIGURE 14 Linear frequency experiments showing how a reduction in driven frequency affects the screen's motion (16 Hz down to 6.3 Hz)

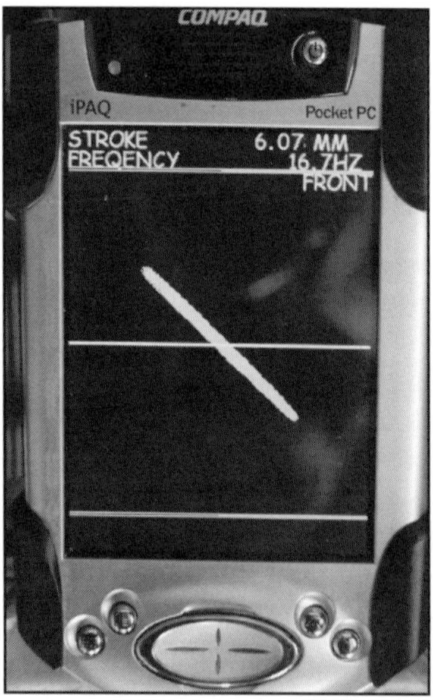

FIGURE 15 Normal screen operation

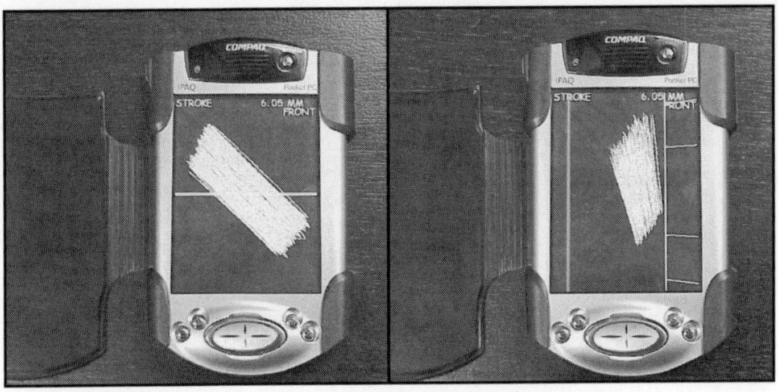

FIGURE 16 Weight on front left tab from xy and zy directions

The screens were examined at the feed end and in the middle. The resulting mean squares of the errors for linear regression (MSE) of axes x and y were examined against the mean squares of the errors for multilinear regression (MSE2) of the three axes: x, y, and z. Differences greater than 0.03 between MSE and MSE2 tended to indicate that the screens may have a motion problem. These problems could include incorrect driven frequency, poor feed distribution, or loose structures.

The motions of four of the screens with differences greater than 0.03 are shown in Figure 17. The fifth 3-D graph shows normal screen motion.

This type of analysis illustrates the ease with which problem screens can be identified; the motion can be observed within seconds of collecting the data and the differences between the MSE and the MSE2 can be determined to aid in the identification of problem screens.

Mass Feed Rate

The stroke was found to change with mass rate to the screen, with the stroke decreasing as mass feed rate to the screen increased. Experiments were conducted at two plants. Both plants had weightometers on the feed conveyors to single desliming screens. This enabled the feed rate to be compared with the stroke, as measured with the portable screen motion analyzer and a fixed triaxial accelerometer package. Figure 18 shows the results of varying the feed rate to the plant and measuring the stroke for 8- and 120-second periods with the portable screen motion analyzer. The longer periods produced acceptable error bars when compared with the 8-second measurement times.

On-line monitoring of a Hunter Valley screen as the tonnages to the plant were varied is shown in Figure 19. These data were collected using accelerometers hardwired to a banana screen, with data analyzed during a 120-second

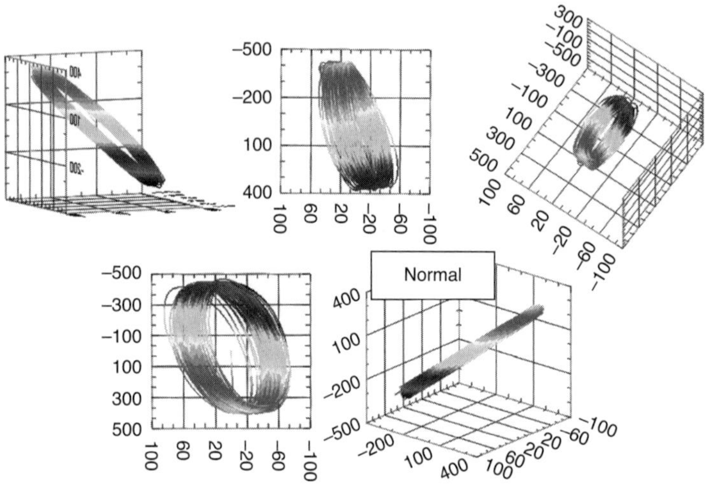

FIGURE 17 Screen motions of screens with differences greater than 0.03 and normal screen motion

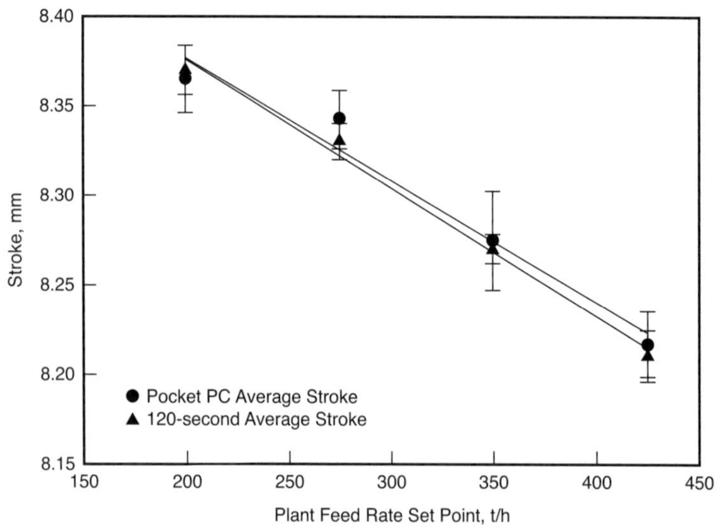

FIGURE 18 Average stroke data from 8-second pocket PC data and 120-second pocket PC data

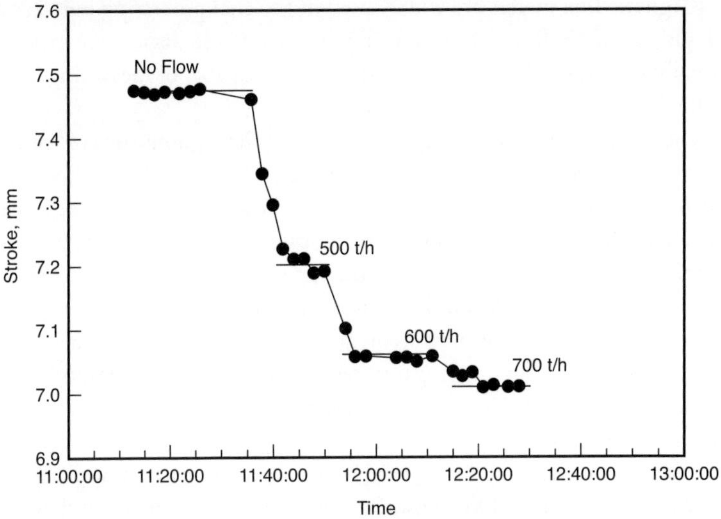

FIGURE 19 Stroke changes with different tonnage rates on-line

period. This plot shows the stroke sensitivity to tonnage decreasing with increasing tonnes to the plant.

CROSS-CORRELATION FLOWMETER

The measurement of flow rates in process streams of coal preparation plants is highly desirable from a process control viewpoint. However, on-line flow rates are rarely measured in most contemporary coal preparation plants because of the difficulties and cost of accurate monitoring. In fact, constant flow rates are provided by fixed-speed pumps and constant sump levels. Nevertheless, variable flow rates can result from variations in solid concentrations, wear in pumps or process equipment, and pump cavitation, leading to sudden loss of efficiency during operation at low sump levels.

Although a number of techniques can be used to measure flow rates in clean fluids, only a few are capable of measuring slurries containing abrasive particles. These include venturi meters, pipe elbow meters, magnetic flux meters, and ultrasonic flowmeters.

Venturi meters and elbow meters correlate pressure drop with average slurry velocity and, if made from suitable wear-resistant material, are suitable for use in abrasive slurry environments. A major disadvantage of these two devices is the need to frequently flush the pressure tapping points to prevent blockage.

Magnetic flux meters are widely used in the coal processing industry and are highly accurate flow rate measuring devices. These devices are expensive, especially for larger pipe diameters, and are not capable of measuring magnetite slurries.

The Doppler flowmeter is also widely used throughout the coal preparation industry. These meters are nonintrusive and can be used as portable or fixed meters. The major problem with these flowmeters is that they need a continuous solid path to the liquid to be measured. Fluid velocity in lined pipes therefore can be difficult to measure, particularly if there is a minor air gap between the pipe wall and the lining material.

Methods of flow-rate measurement based on the transit time of some kinds of disturbance, such as electrical resistance between two points in a flow system, provide alternative approaches for the flow-rate measurement in coal preparation processes. The basic theory of the methods was developed in the 1950s by Taylor (1953). Beck (1981) described the earliest application of transit time method in slurry flow. In their work, solids in slurry flows were used as naturally occurring tracers, and the transit time was measured using hardware-based cross-correlation. Because cross-correlation (or transit time) flowmeters developed in the 1970s and 1980s were based on electrical hardware for signal measurements and cross-correlation, the cost of the flowmeters was prohibitively high. Although the method was technically successful, developments in other flowmeters at lower costs meant that cross-correlation flowmeters were not a cost-effective alternative in the 1970s and 1980s. However, the availability of cheap and high-performance computers and digital microelectronic systems in recent years can significantly reduce the cost of cross-correlation flowmeters based on digital techniques.

A schematic representation of the cross-correlation flowmeter technique is shown in Figure 20. $X(t)$ and $Y(t)$are alternating current (ac) electrical signals measured at upstream (X) and downstream (Y) sensor locations. $R_{xy}(\tau)$ is the cross-correlation of the two signals, given by

$$R_{xy}(\tau) = \frac{1}{T}\int_0^T X(t)Y(t-\tau)dt \qquad \text{(EQ 2)}$$

where T is the total sampling time and τ is the transit time of similar signal perturbations.

The mean transition time, τ^* of resistivity fluctuations between two electrode pairs separated with a distance L is given by

$$\tau^* = \frac{1}{T}\int_0^T R_{xy}(t)dt \qquad \text{(EQ 3)}$$

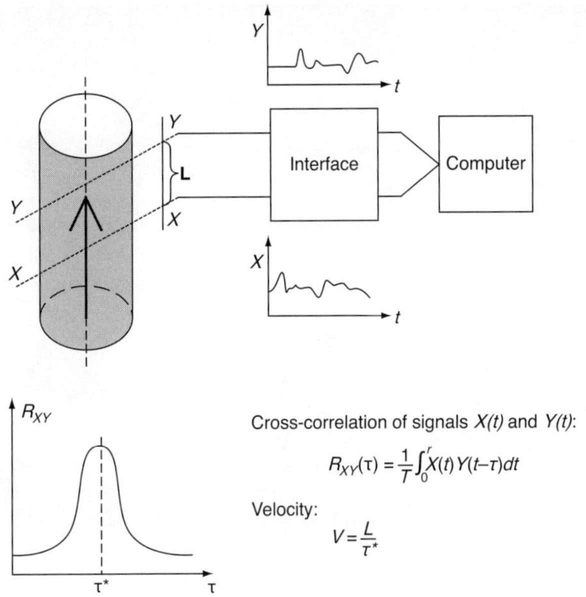

FIGURE 20 Schematic representation of the cross-correlation method

The mean velocity V is therefore given by

$$V = L/\tau^* \tag{EQ 4}$$

A new cross-correlation flowmeter has been developed by combining digital microelectronics and software for signal measurement and analysis. The digital cross-correlation method is used to calculate the transit time of resistivity fluctuations accompanying slurry composition fluctuations between two electrode pairs of known separation so that velocities can be directly calculated. Precise and safe measurement of ac electrical signals in the voltage range of 1–20 V requires simple instrumentation. Thus, the flow-rate measuring technique is highly reliable, inexpensive, and applicable to multiplexing.

Hardware

Hardware circuits consist of signal generation, demodulation, electrodes, and tuning circuits. The sensor probe (Figure 21) consists of 316 stainless-steel electrodes 0.9 mm in diameter, spaced 3 mm apart. The electrodes are cast in a wear-resistant epoxy to minimize abrasive wear. The wear-resistant epoxy was the most convenient at the time for its ease of use, but much greater strength and durability can be obtained with the use of ceramics or basalt as a cast

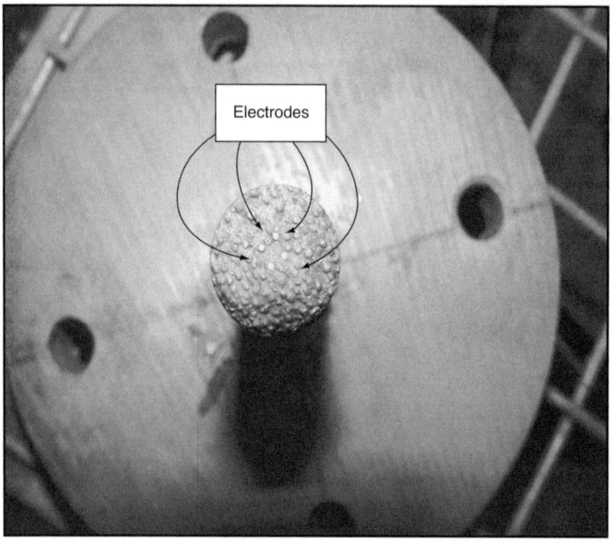

FIGURE 21 Sensor tip showing electrode configuration and wear-resistant epoxy

material because it was found that the organic matrix of the wear-resistant epoxy wore preferentially, leaving the harder particles standing proud of the surface. This could give rise to the formation of eddy currents around the electrode. A flange was attached to the sensor (Figure 22), and the sensor was designed to be inserted into a 20-mm ID (inside diameter) ceramic-lined flanged port built into a specially constructed basalt-lined spool piece. The spool piece was inserted into the dense-medium cyclone feed line, replacing an existing pipe section. Sensor length was adjustable so that the electrode tips could be aligned flush with the internal wall of the basalt-lined pipe section. In practice the tip was slightly proud of the surface, resulting in a higher-than-expected wear rate.

Software

LabView software was used to filter, cross-correlate, and calculate the slurry velocity. This software was chosen because of its adaptability and ease of use in the field. In addition, the program can be compiled for distribution to Windows-based computers.

The signals from two measurement channels were passed through the cross-correlation function to determine the time of flight between perturbations. The distance between the electrodes (3 mm) was then divided by this time to give the velocity of the slurry passing the electrodes.

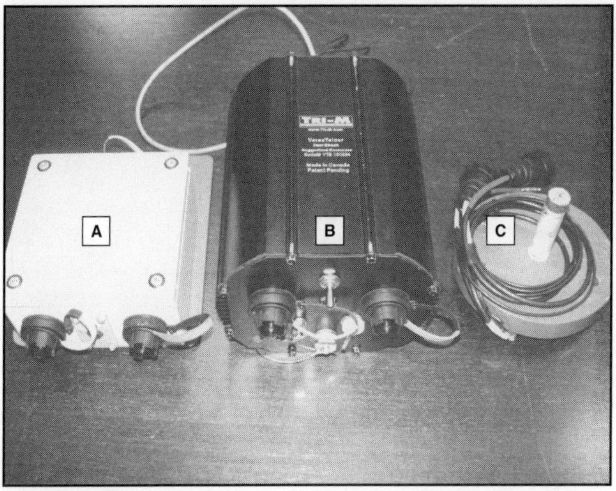

FIGURE 22 Complete cross-correlation instrument: (a) tuning circuit, (b) container incorporating demodulator and signal generator, and (c) flanged sensor unit

Validation

Initial validation was performed in the pilot plant using pipes with diameters of 40 mm and 150 mm. The distance between the probes varied from 3 mm to 50 mm, and solid contents varied between 0.3% and 60% solids (by weight), using coal, sand, and flotation tailings as solids. The system was also tested using magnetite and magnetite–coal slurries at typical concentrations found in coal preparation plants with similar results. Figure 23 shows the results of the cross-correlation flowmeter and a magnetic flowmeter installed in the same pipe. The large variation shown by the error bars is the result of variations in the conditions under which the measurements were taken, such as electrode spacing and software filter settings.

Plant Experiments

Operators of a Bowen Basin plant agreed to host a trial of the cross-correlation flowmeter. A spool piece was manufactured to replace a section of the feed pipe to one of the dense-medium cyclones. The 300-mm-ID basalt-lined spool piece was equipped with a 20-mm ceramic-lined port and flange for the cross-correlation sensor, as shown in Figure 24.

The plant experiments were carried out in two stages, 2 months apart. The probe was left in the pipe for the 2 months before the second stage, which provided the opportunity to determine wear and the effect of wear on the probe signal.

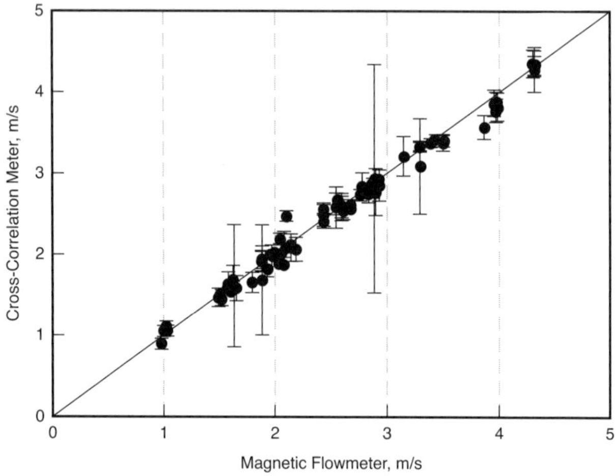

FIGURE 23 Plot of all data recorded for cross-correlation flowmeter compared with data from magnetic flowmeter

FIGURE 24 300-mm ID basalt-lined spool piece inserted into dense-medium cyclone feed line showing flanged port, inserted sensor, and tuning electronics

Fitting of the flanged electrode occurred during a planned maintenance shutdown, and readings were taken the next day, while the plant was operating in normal production mode. Attempts were made to measure the flow with a portable Doppler meter; however, no suitable signals could be measured

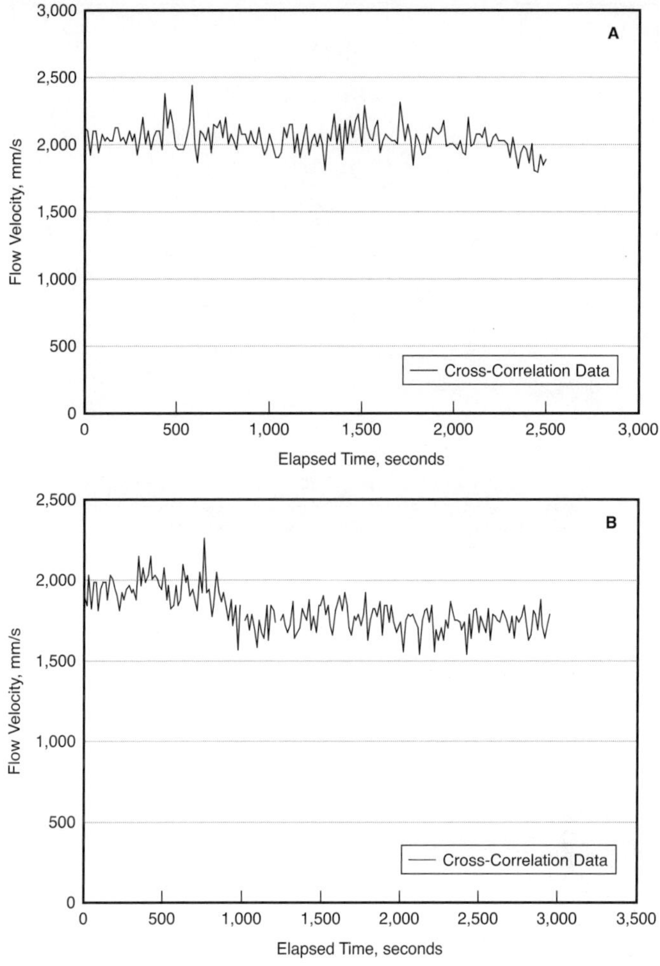

FIGURE 25 Cross-correlation meter readings from June 11, 2004: (a) 1:49–4:37 PM, and (b) 4:45–5:49 PM

through the basalt lining–steel interface. Flowmeter readings were recorded over a 2-day period while analysis parameters such as filter conditions, count rates, and total analysis times were varied. It took approximately 6 seconds to record and process the data for the 3-second period used. Recorded flows for the second day are shown in Figure 25. Both graphs show a variation in the measured flows of approximately ±5% on average, with variations as high as ±10%. This variation was unexpected, with obvious implications for the operation of the dense-medium cyclone. Unfortunately, over the time period in

FIGURE 26 Close-up of probe tip showing wear marks

which these recordings were taken, the pressure gauge to that particular dense-medium cyclone was not operational and pressure measurements were not available. As previously stated, attempts to measure the flow using a portable Doppler meter failed because of the basalt lining–pipe interface problems; however, the velocities recorded (approximately 2 m/s) were realistic for this system.

During the second trial 2 months later, data were collected from both the cross-correlation flowmeter and the pressure sensor to the dense-medium cyclone. The cross-correlation sensor had been left in the pipe from the previous trip in June, but no opportunity to inspect the probe tip was available before flow measurements were taken.

The probe was removed at the end of the second trial period, and Figure 26 shows a close-up of the probe tip. The direction of flow is shown by the surface wear (from bottom left to top right). The probe tip is worn most on the edge that was presented first to the flow stream (bottom left to middle of Figure 26), which may indicate that the sensor was extended slightly into the flow stream and not flush with the pipe lining. The actual probe electrodes have

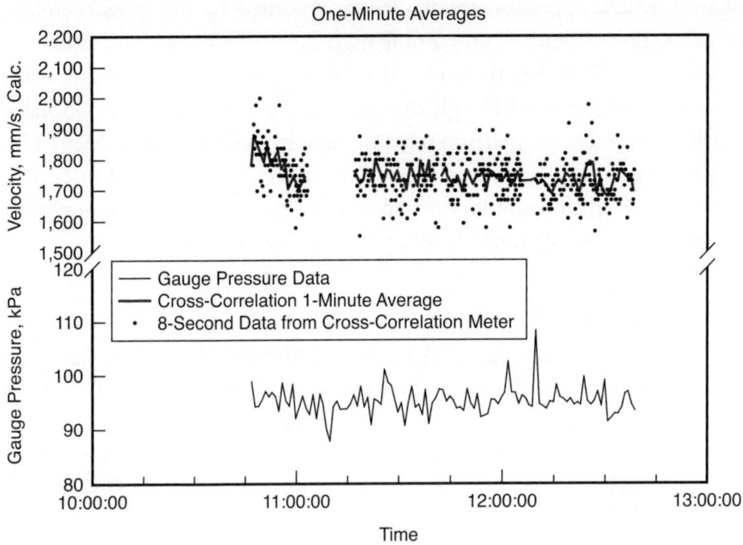

FIGURE 27 Plot of cross-correlation meter output and pressure

also been worn and are slightly recessed compared with the wear-resistant particles, which stand proud from the matrix. Some marks on the surface of the electrodes are also evident.

Cross-correlation data and pressures from the dense-medium cyclone under test were collected, and some of these data are shown in Figure 27. Whereas the cross-correlation data were collected at 8-second intervals, only 1-minute averages of cyclone pressure data were available. One-minute averages for the cross-correlation data were calculated and plotted in Figure 27 for comparison.

Variations in the pressure data in the dense-medium cyclone were of similar magnitude to those of the cross-correlation meter data (Figure 27), although no correlation was found between the two. The cross-correlation meter data varied by as much as 10% during this trial, as shown by the 8-second data plotted in Figure 27. This variation was twice that of the previous trial and may be a result of the lower signal strength caused by probe wear. This variation is a new finding and warrants a more focused investigation before its importance and the need for corrective action are fully understood.

No direct measurement of slurry velocity was available for comparison with the cross-correlation flowmeter output. However, the flow rate calculated from the nominal feed tonnage to the dense-medium cyclone (190 t/h), assuming a 3-to-1 medium to coal ratio and a solids density of 1,600 kg/m^3, is

1.86 m/s, which approximates the values recorded by the cross-correlation flowmeter. The difficulty in obtaining high-quality flow-rate data to compare with the cross-correlation flowmeter data highlights the problem of measuring slurries in lined pipes. Although the data can be validated in the laboratory, the lack of a suitable comparison measurement in the plant is a problem. However, this technique is a primary measurement because it does not infer or calculate the flow from secondary properties or from a calibration equation but relies on primary measurements of time over a fixed distance.

COMMERCIALIZATION

CSIRO Energy Technology and Ludowici Mineral Processing Equipment signed a commercial agreement for these instruments in August 2006 and have the option to further develop other instrumentation techniques currently being developed by staff of CSIRO Energy Technology's coal beneficiation group at QCAT.

ACKNOWLEDGMENTS

The financial support of ACARP is gratefully acknowledged. The authors would also like to thank the ACARP industry monitors and the operators and staff of the plants during the trial of these instruments.

REFERENCES

AS 1038.24-1998. *Coal and Coke—Analysis and Testing: Guide to the Evaluation of Measurements Made by On-Line Coal Analysers*. Sydney, NSW: Standards Australia.

Beck, M. 1981. Correlation in instruments: Cross correlation flowmeters. *J. Phys. E Sci. Instrum.* 14:7–19.

Firth, B., O'Brien, M., Ofori, P., Jensen, B., He, Y., Stanmore, B., and White, E. 1995. Use of an on-line moisture meter to investigate product moisture. *AusIMM Proc.* 2:47–51.

Grubbs, F.E. 1948. On estimating precision of measuring instruments and product variability. *J. Am. Stat. Assoc.* 43:243–264.

———. 1982. An introduction to some precision and accuracy of measurement problems. *J. Test. Eval.* 10:133–143.

Myers, A., Leach, R., Clarkson, C., Edward, D., Lahey, A., Firth, B., and O'Brien, M. 2000. *Performance of Banana Screens in Drain and Rinse Applications*. ACARP Project C7048. Brisbane, Australia: Australian Coal Research.

O'Brien, M., Firth, B., Myers, A., Nemeth, D., and Burger, A. 2002. Improved Performance of Banana Screens in Drain and Rinse. ACARP Project C8042. Brisbane, Australia: Australian Coal Research.

O'Brien, M., Hutchinson, T., and Wesner, C. 2003a. *Portable Screen Motion Analyser*. ACARP Project C11005. Brisbane, Australia: Australian Coal Research.

O'Brien, M.E., Nemeth, D., Wex, T., and Sullivan, S. 2003b. *Image Analysis Measurement of Screen Aperture and Open Area*. ACARP Project C11006. Brisbane, Australia: Australian Coal Research.

Taylor, G.I. 1953. Dispersion of soluble matter in solvent flowing slowly through a tube. *Proc. R. Soc. London, Ser. A* 219(1137):186–203.

SUPPLEMENTAL REFERENCES

O'Brien, M., Hu, S., and Hutchinson, I. 2002. *A New Flowrate Measuring Technique for the On-Line Monitoring of Coal Washing Processes*. ACARP Project C10046. Brisbane, Australia: Australian Coal Research.

———. 2005. *Cross Correlation Flowmeter Plant Trial.* ACARP Project C10046, Stage 2. Brisbane, Australia: Australian Coal Research.

Building in Preventive Maintenance

Greg DeHart

ABSTRACT

A well-maintained coal processing plant greatly affects the profitability of an operation. Because of this, much forethought should be applied to the maintenance program during plant design and construction. This chapter describes the basic design philosophy and considers each of the major components and how design can impact maintenance. The final section on predictive and preventive maintenance discusses how to maximize both availability and component utilization. Because of many advances in materials and predictive technologies, optimum plant performance can be achieved through utilization of these tools in a good maintenance program.

INTRODUCTION

The coal industry has been forced through market pressures to continue to create ways to produce coal at lower cost. The coal preparation side of production has made tremendous strides with innovative equipment, materials, and techniques to meet the demand. These developments have led to more reliable plants that are capable of functioning longer hours between maintenance intervals. This has given the industry more confidence to decrease feed tonnages because plants operate more hours, lowering capital needs. The continuing decrease in the number of people interested in working in the industry has led to a decrease in the workforce skill levels. Plants are quickly adapting by trying to reduce the skills needed to maintain plants. Therefore, the coal preparation industry is being forced to become more cost effective, safe, and reliable.

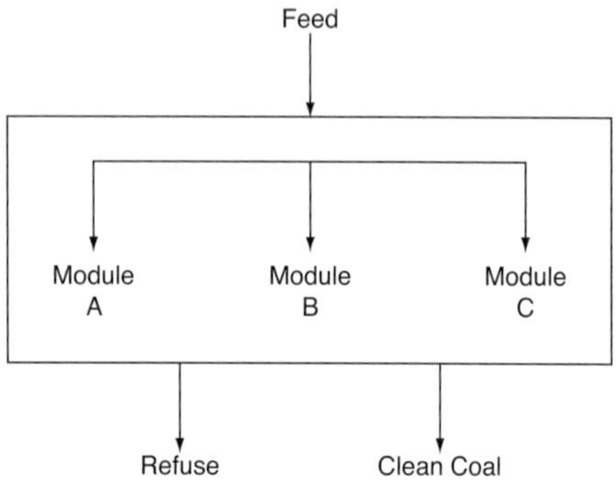

FIGURE 1 Multiple circuit layout

PLANT DESIGN PHILOSOPHY

The continuing innovations in process equipment have led to a major reduction in the unit numbers needed to process a ton of coal. The introduction of the multislope vibrating screen, also known as the banana screen, has led the way in this area. Because of the screen's ability to process much greater capacities in comparison to horizontal machines, fewer screening units must be maintained. Other process equipment has evolved to match the larger-capacity abilities of these screens, leading to plant designs that use single-unit machines in each stage of coal processing. This innovation has eliminated tremendous amounts of square footage in plants, so fewer chutes, pipes, and other ancillary supports are needed.

Circuit capacities of 800 tph or lower facilitate current standards of proven and reliable equipment design. This capacity seems to be the benchmark for wide-body screens that have proven reliability (i.e., 10 ft × 24 ft maximum). Use of a combined raw coal–desliming screen eliminates both floor space and elevation. Use of a combined dilute media recovery circuit allows the elimination of a pumping circuit and associated piping. If more capacity is needed, multiple circuits not exceeding the 800-tph limit per circuit (Figure 1) can be used. Multiple circuits also facilitate maintenance without interrupting coal production.

DESIGN OF PLANT COMPONENTS

Chutes

From the beginning design stage, it is critical to consider how material will move through the plant. Several factors dictate the design of the chute work throughout the plant. Issues include material type, size, fall, and velocity. In order to allow for all circumstances, a 20% overcapacity factor should be applied to design calculations. Material sliding through the chutes creates most of the wear. In order to extend chute life, the plant should be designed to minimize or eliminate directional changes, which lead to impact damage. Properly designed ceramic chutes have the benefits of longevity, isolation of a particular skill, and plant cleanliness. Although ceramic linings double the cost of chutes, the improved performance through reduced maintenance makes the system cost effective. The ceramic can easily be replaced in the field by a specialty company technician, eliminating the need for skilled metal fabricators and frequent liner replacement. Reduction in specific skilled staff reduces plant cost.

Screens

The design philosophy of single-unit operations leads to the use of large screens to handle all necessary material for each circuit. Accurate and efficient sizing of material is critical to proper performance in the separating circuits. Therefore, screening surfaces are used, maximizing wear life while still meeting efficiency requirements. Each operation must consider the characteristics of the plant material to determine which screening media will be applicable in their particular situation.

Two recent advancements in the area of screen design related to screen maintenance are the use of wear-resistant materials and modular design. The use of ceramic lining has extended the life of cross-members, feed boxes, and discharge lips. The added expense is offset by the reduced maintenance and improved availability of the equipment. The wear resistance of the screening media, while maintaining an efficient separation, has been improved through the use of urethane decks and chrome plating of stainless-steel profile wire. The other improvement in screen design relates to the screen decks. The construction of the decks with a modular bolt-on design allows replacement of certain parts as they wear without needing to replace the entire deck.

Dense-Medium Vessels

Vessel maintenance has been facilitated by the development of new materials and drive technologies. The use of larger chains and chromium carbide in the vessel construction allows longer intervals between major unit overhauls without the danger of on-shift failure. In addition, the use of fluid coupling and variable frequency drives has led to improved reliability and lower component costs.

Centrifugal Dryers

The large-capacity, single-unit design method has been carried through to centrifugal dryers. Having only one unit per circuit greatly simplifies the maintenance program. Using a dryer design that keeps the components of the dryer above the floor line facilitates maintenance and repair.

Pumps and Piping

In the design of a pumping system, the highest efficiency rating possible should be used. Operating the pump in the proper range leads to good pump life. In addition, the use of 28% chrome as a minimum for all wear components delivers a highly efficient and robust pump.

All process piping is engineered to optimize the use of pumping power and limit the amount of pipe wear. The piping material used for slurry carrying particles greater than 1 mm should have ceramic-lined elbows, with basalt piping or ceramics for long, straight runs. High-density polyethylene (HDPE) piping can be used for pumping slurries containing particles less than 1 mm. In this application, the elbows and fittings should be ceramic (Figure 2). For elbow discharges, a straight run of ceramic pipe should be installed that is at least twice the radius of the elbow. For example, for a 90° elbow with a radius of 2 ft, a 4-ft length of pipe should be used on the discharge. This is done to protect the pipe from wear that would be caused by a rotating spray discharging from the elbow.

FIGURE 2 HDPE pipe with ceramic elbow and extension

All piping should be engineered pipe with bolted flanges and should be well identified for future maintenance. Experience has shown that the use of these improved materials will result in a pipe life of four to eight times that of carbon steel piping. The *Handbook of Polyethylene Pipe* (Plastics Pipe Institute 2006) states that there should be an increase in pipe life approximately three to five times that of steel. According to the handbook, the difference in estimated life comes from the difference in slurry velocity of 15 ft/s and the plant velocities, which are more commonly 8–10 ft/s. Plant velocities should be lower than that, so pipe life should be greater. The use of flanged pipe coupled with the extended pipe life removes the need for a skilled pipe fitter to be on staff to repair and replace steel pipe. Having reliable and well-maintained piping systems keeps leaks and sprays from worn pipes to a minimum.

Conveying Systems

In the design of the conveying systems, Conveyor Equipment Manufacturers Association (CEMA) standards should be applied, with special attention to transfer points and belt-wiping systems. The key to maintaining a properly designed conveying system is the cleanliness of the system, which reduces wear and facilitates access to components. The use of good belt wiping and effective skirting keeps the material in the proper location. It is also important for the design to be able to handle worst-case scenarios, and investment in the planning stages can prevent problems when extreme conditions occur.

Facilitating Maintenance

Access to units and components for maintenance is crucial to the overall effectiveness of a maintenance plan (Figure 3). The installation of bidirectional hoisting systems can greatly reduce the labor involved in maintaining large equipment and can lead to a safer work environment (Figures 4 and 5). Although consideration of overhead space is important, such systems can also help plan for spills and other materials that inevitably use floor space. Drains and cleanout boxes throughout the plant will allow the staff to keep the plant clean and safe.

PREDICTIVE AND PREVENTIVE MAINTENANCE

With the complexity of the modern coal preparation plant and the lower number of employees, a computer-based system must be implemented to track the condition of equipment. This position must be staffed properly to ensure that the full benefit of preventive and predictive maintenance is realized. This strategy involves state-of-the-art monitoring technologies such as oil analysis,

FIGURE 3 Example of open floor plan to allow access from overhead crane

FIGURE 4 Overhead crane

FIGURE 5 Articulating hoist beam for pump maintenance

thermography, vibration, ultrasound, and motor circuit monitoring. The effectiveness and implementation of these programs has been discussed by Van de Ven (2006).

It has been said that where there's smoke, there's fire; and in the plant environment, where there's heat, there's friction. Simply stated, friction leads to wear and failure. A remote temperature monitor, or heat gun, is one example of an inspection tool that plant employees can use on a regular basis to establish baselines and prevent failure of critical components. The employees are central to the success of this program. Educating and providing them with up-to-date information will improve maintenance. If the employees take personal pride and ownership in their work, it will be reflected in the operation and availability of the plant. Through proper preventive, predictive maintenance and condition monitoring, the largest percentage of useful life can be realized.

Although time can be an indicator of failure, it is not usually the best gauge. If equipment is replaced based on the condition of critical components instead of the operating hours, the parts will be fully used (Allied Reliability

2007). In addition, because of the tight limitations of technical proficiency in the coal industry today, a close relationship with equipment manufacturers is vital. This philosophy is reflected in service contracts with manufacturers and the use of original equipment manufacturer parts in all repairs.

SUMMARY

From the beginning of the design and construction of the modern coal processing plant, maintenance is at the forefront of planning. From the selection of proper equipment and layout of the plant, many decisions must be made, which are reflected in a well-maintained and well-operated plant. Market pressures have forced the coal production industry to respond by doing more with less. This has been accomplished through the use of new materials and technology along with fewer employees, especially fewer specialized employees. Maintenance must be built into the plant structure, into the equipment, and, possibly most importantly, into the culture.

REFERENCES

Allied Reliability. 2007 *Are You Doing Too Much PM?* www.alliedreliability.com/admin/files2/files/PMReport.pdf. Accessed August 2007.

Plastics Pipe Institute. 2006. *Handbook of Polyethylene Pipe*. Irving, TX: Plastics Pipe Institute. p. 80.

Van de Ven, H. 2006. Stop walking through the maintenance minefield with your eyes closed. *CPSA J.*, 5(3-4):19–20, 15–17.

Status of Current Coal Preparation Research

R.Q. Honaker, G.H. Luttrell, and P. Bethell

ABSTRACT

The focus of recent coal preparation research reflects the needs of the industry to optimize operations to achieve maximum profitability while meeting contractual quality specifications. In many cases, the current fine coal processing technologies do not provide the necessary performance to achieve a true optimum yield condition. Recent research has addressed the development of new fine particle devices for density and particle size separations. Dewatering has remained a significant issue, with the need to decrease product moisture levels while maximizing particle recovery. Where moisture levels are critical and water supply is limited, dry coal cleaning is being implemented, which has rejuvenated the research effort in this area. This chapter reviews the status of coal preparation and the recent and ongoing research that supports current industrial trends.

INTRODUCTION

Research has played an important role in developing the technologies used in today's coal preparation plants. Throughout the early 20th century, a significant number of dry and water-based separation technologies were developed, studied, patented, and commercially implemented, including several dense-medium separation technologies and froth flotation. The research conducted by the Dutch State Mines in the 1940s and 1950s contributed greatly to the development and operating knowledge of dense-medium cyclone (DMC) circuits and is still used today in conjunction with other significant contributions such as those from the U.S. Bureau of Mines and the Julius Kruttschnitt Mineral Research Centre. In a majority of coal preparation plants, fine coal recovery involves flotation technologies that were developed from more than 100 years of research. The most recent development is the column flotation cell, which was derived from research conducted in the past three decades. During this same period, many other innovative technologies have evolved from extensive research programs and have been commercially implemented to improve the overall efficiencies of coal preparation.

Most coal cleaning facilities use water as the medium to upgrade coal quality. In comparison to dry cleaning separators, water-based cleaning technologies provide superior efficiencies at the separation density values typically needed to achieve the required clean coal product qualities. Although many variations in coal cleaning plants exist, a generic flowsheet can be envisioned for a typical plant based on current trends in the design of new plants and the renovation and expansion of older facilities.

In the United States, most plants are forced to operate at a very high level of processing efficiency to compete in the domestic market. This objective typically necessitates the use of several parallel processing circuits, often three or more, that are specifically designed to optimize the processing of each size fraction. Many plants also justify the use of a separate coarse cleaning circuit to avoid crushing and to minimize the tonnage of coal that is processed in less efficient fine circuits. The washability characteristics of typical U.S. feed coals make it difficult to justify size reduction to improve liberation. Furthermore, the strict limits on moisture imposed in U.S. markets provide a strong incentive for plant designers to keep the feed size as coarse as possible via the use of a coarse cleaning circuit. As a result, a modern flowsheet for a U.S. plant would typically include four independent circuits for treating coarse (>10 mm), medium (1–10 mm), small (1–0.15 mm), and fine (<0.15 mm) coal (Bethell 2002; Bethell and DeHart 2006; Holcomb et al. 2006). Figure 1 shows a simplified layout for such a plant configuration. The coarse fraction is typically treated using a chain-and-flight dense-medium vessel, whereas the medium fraction is upgraded using one or more DMCs. The small coal fraction is typically treated by water-only cyclones, spirals, or a combination of these separators. The finest size fraction may be treated directly by froth flotation, deslimed to remove a substantial portion of minus 35–40 µm solids before flotation, or discarded as waste in some older plants. Dewatering is typically performed using centrifugal basket-type dryers for the coarser factions and screen-bowl centrifuges for the finer fractions. Vacuum filters are used in a few plants but are difficult to justify unless a thermal dryer is available to ensure that moisture specifications can be met.

Another design philosophy, often used in Australia, focuses on simplification and robustness, which often are attained through the use of fewer processing circuits. Two distinct flowsheets, each with two parallel circuits, can be identified that operate in either the metallurgical or steam coal market. A coarse coal circuit is notably absent in both cases. The top size of the feed is typically smaller than that of the plant described in Figure 1 at around 50 mm,

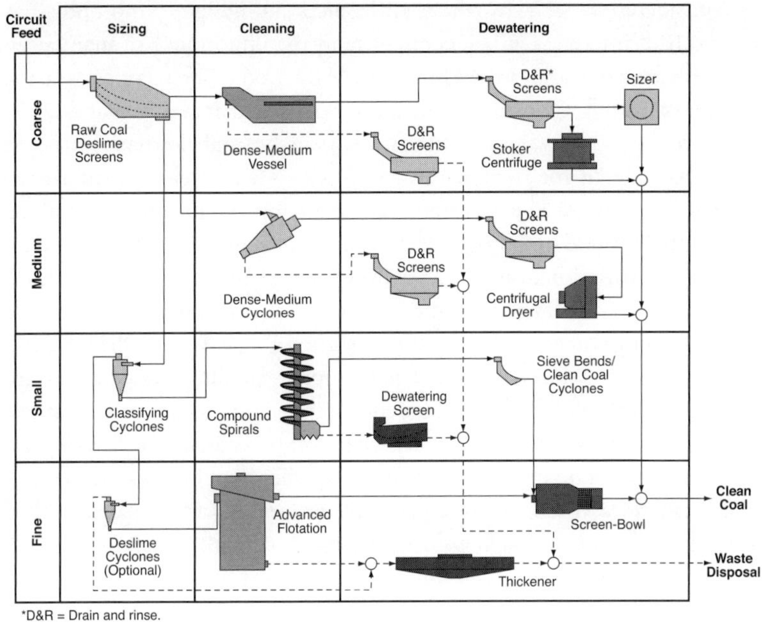

FIGURE 1 Conceptual flowsheet for a typical wet-based preparation plant

which is a common top size for utility coal and improves liberation of mineral matter. The liberation advantage is important for coal that contains a significant amount of middling particles but could lead to an increased production of fine coal. For metallurgical applications, a typical flowsheet would include DMCs to upgrade coal particles in the nominal 50–0.5 mm size range. The small and fine coal not treated by DMCs is typically processed using froth flotation circuits. This circuit design is inherently inefficient because of the misplacement of coarser coal into the flotation circuits. Centrifugal dryers are generally used to dewater the dense-medium product, whereas vacuum disc filters or, more recently, horizontal belt vacuum filters dewater the small and fine coal. A different layout specifically addresses the steam coal market. Whereas the same basic configuration of DMCs is used to treat solids nominally coarser than about 1 mm, water-based separators (such as spirals or teeter-bed units) are used to treat the small and fine coal coarser than about 0.10–0.15 mm. Both of the clean coal products are dewatered using various configurations of centrifugal dryers. Because no attempt is generally made to recover the finest size fraction of coal, this material typically is discarded as waste from the classifying cyclone circuit.

Consideration of why these differences in design philosophy have emerged is of interest. Designers understand the importance of maintaining good probable error (Ep) values for each size fraction treated. For sizes coarser than approximately 1 mm, dense-medium processes such as vessels and DMCs provide very good Ep values. On the other hand, several alternative processes may be considered for finer sizes such as water-only cyclones, spirals, and teeter-bed separators. The selection of the most appropriate unit must address not only its Ep value but also the ability of the unit operation to maintain an incremental quality that can optimize the overall plant. It is widely recognized that plant yield is optimized when the same incremental quality is maintained in all processing circuits (Cierpisz and Gottfried 1977; Abott 1981). Furthermore, each size fraction treated in a given process should ideally be separated at the same incremental quality. This additional requirement is often difficult to attain for some water-based density separators, which in most cases separate larger particles at a much lower incremental quality than finer particles.

Recent research efforts have generally focused on areas that directly or indirectly address the design principles described in this section. Also, the need for dry coal cleaning technologies is increasing, which has resulted in a significant amount of research effort in modifying technologies developed in the early 20th century and developing new technologies that may lead to dry coal cleaning plants that upgrade the coal in all size fractions. This chapter will discuss a few recent research developments in both the water-based and dry cleaning areas.

WATER-BASED COAL CLEANING RESEARCH

On-Line Efficiency Analysis

In separate studies, researchers at Virginia Tech (Blacksburg, Virginia) and the Council for Science and Industrial Research (CSIR) in South Africa have developed the use of transponder technology for on-line monitoring of density separation efficiency in coal preparation (De Korte et al. 2002; De Korte 2003). Currently, on-line efficiency studies are performed by injecting a number of polymer-based blocks into the feed stream of a gravity-based separator. The blocks are commonly called tracers and have a range of known solid densities due to varying compositions of a fill material, such as magnetite. Employees are placed at the exit points of the product and tailing streams to handpick the tracers from the process streams. The efficiency information is then derived from the recovered tracers. Obviously, the success of the efficiency study is largely reflective of the number of tracers recovered.

The researchers at Virginia Tech and CSIR placed a small radio transducer into each of the tracers, which are called Supertracers. As a result, the tracers are easily detected by a radio-frequency antenna placed at the exit point of each process stream. Thus, errors in the efficiency analysis caused by unrecovered tracers are eliminated.

Density separators were evaluated at two preparation plants using the Supertracer technology. At the Tavistock plant in South Africa, the study involved the use of 23-mm tracers to measure the efficiency of Larcodem units. The Ep value determined by the Supertracers (0.011) compares well to the value obtained by the conventional float-and-sink method (0.007) (De Korte et al. 2002). A similar favorable comparison was achieved at the Goedehoop preparation plant.

Fine Density-Based Separations

A circuit with large-diameter DMCs, low feed pressures, and conventionally sized magnetite has been investigated for cleaning coal having a particle size below 1 mm in a pilot-scale circuit having a throughput of 25 tph (De Korte 2002). Using a single-stage DMC, these conditions provide a poor separation efficiency with a significant loss of coal. However, using a rougher-scavenger-cleaner circuit, a separation density of around 1.8 was achieved in the initial evaluations, with an Ep value of around 0.066, which yielded an organic efficiency of 98.5%. The magnetite consumption was determined to be 1.5 kg/t of coal treated.

Teeter-bed separators have been installed in several coal preparation plants in Australia and the United States as an alternative to spiral concentrators. The benefit is that the units have very high mass flow capacities, in the range of 20–40 tph/m^2, which eliminates or minimizes negative impacts of poor feed distribution that can occur with spiral concentrator circuits.

Two major development and demonstration projects were recently completed on a pair of novel teeter-bed separators. The CrossFlow separator, which is commercially represented by Eriez Manufacturing, eliminates the negative impact of the feed being injected into the center of the teeter-bed, as is done with conventional units. The feed to the CrossFlow separator enters from one side of the unit, and particles that enter the fluidized bed must first settle out of the film layer that is flowing across the unit, as shown in Figure 2a. A demonstration program funded by the U.S. Department of Energy optimized the separation performance of a 2.75 × 2.75 m^2 separator that was designed to treat 2 × 0.25 mm coal at a capacity of 200 tph (Kohmuench et al. 2006). Figure 3 shows the improvement in separation performance from the initial start to the final series of tests. From Test Set 3, the product ash was reduced from around 18% to 10%, and 95% of the combustible material was recovered. The organic efficiency is approximately 97%.

Another method of increasing the throughput capacity of a teeter-bed separator is the use of inclined plates. Researchers at the University of Newcastle

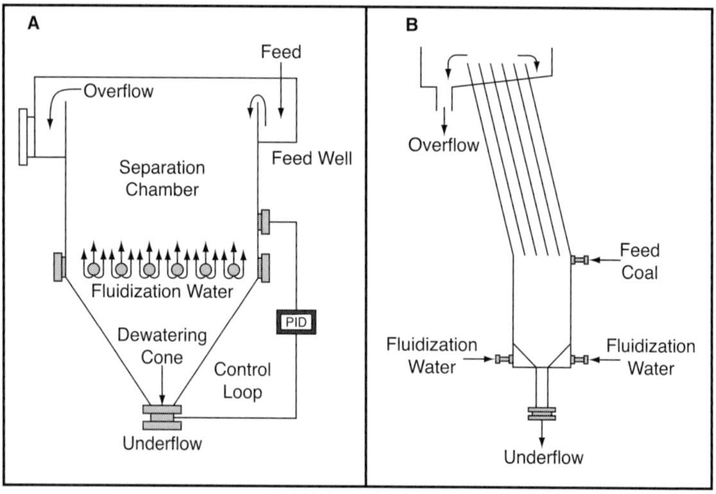

FIGURE 2 Schematics of (a) CrossFlow and (b) reflux teeter-bed separators

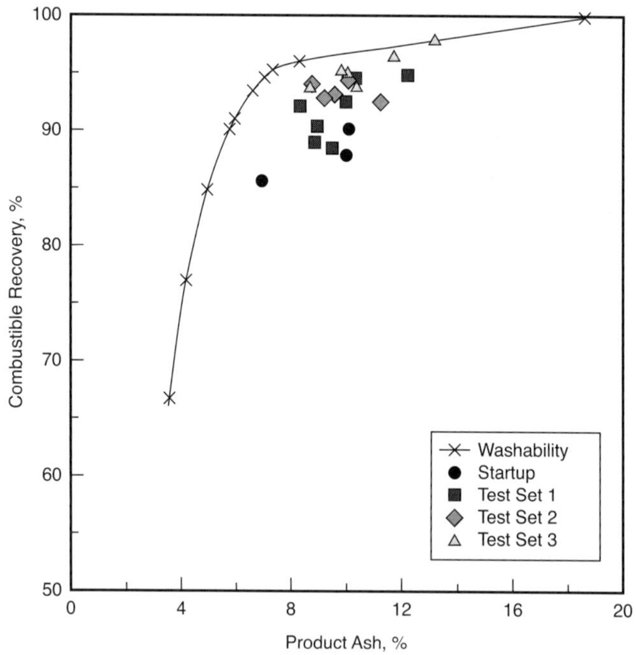

FIGURE 3 Separation performances from a full-scale CrossFlow separator

TABLE 1 Particle separation efficiency achieved by the reflux classifier

Parameter	Overall	Particle Size Fraction, mm				
		2 × 1.4	1.4 × 1.0	1.0 × 0.7	0.7 × 0.5	0.5 × 0.25
ρ_{50}	1.70	1.47	1.53	1.60	1.74	1.91
Ep	0.15	0.04	0.03	0.06	0.08	0.15

(Australia) have developed a teeter-bed unit that uses the accelerated particle movement supplied by the inclined plate concept, as shown in Figure 2b. The unit is commercially known as the reflux classifier and is commercially represented by Ludowici. Separation performances achieved from a test program on a full-scale reflux unit at a feed mass flux of 16 tph/m^2 revealed exceptional efficiency across a particle size range of 2 × 0.25 mm, as shown in Table 1 (Galvin et al. 2004). Because of the upward separation density shift with decreasing particle size, the overall efficiency was lower than any of the size fractions. The overall relative density (RD) of separation (ρ_{50}) of 1.7 at an Ep of 0.15 is comparable to a compound spiral circuit middlings recycle. Interestingly, a new project funded in 2007 is investigating the use of the reflux classifier as a dry coal cleaner in which air is used as the fluidizing medium.

The application of density-based separators for cleaning particles below 0.15 mm in size has been a focus of several research projects in the past two decades, with most of the effort addressing the development of devices using a mechanically applied centrifugal field. However, recent research has indicated that lowering the feed volumetric flow rate and solid concentration to about 50 L/min and 15% solids by weight, respectively, to a compound spiral concentrator can result in a ρ_{50} of 1.8 RD for the 250 × 44 µm size fraction, with an Ep value of 0.20 (Honaker et al. 2006b). From a compound spiral circuit installed to clean nominally 150 × 44 µm coal, the ash content was reduced from 17.35% to 9.84% and the total sulfur from 3.56% to 3.00%.

Classification

The most significant issue that limits the application of density-based separators such as compound spirals and centrifugal separators (i.e., Falcon or Knelson concentrators) to cleaning minus-0.15-mm coal is the inability to remove the high-ash slime material from the concentrator product. Other needs for efficient ultrafine sizing units are in deslime flotation circuits and applications where the +25 µm fraction in the feed has a coal quality sufficient to meet market specifications without the need for a density-based separator or froth flotation unit.

The most direct means of achieving particle size separations is through the use of screening. Low capacity and blinding issues have generally prohibited the use of screening below 300 µm in the coal industry. However, recently developed screening technologies provide the potential to achieve efficient particle size separations below 150 µm at capacities amenable for use in coal preparation plants. Plant data from a recent in-plant study revealed that the SuperStacker technology by Derrick provides a very efficient particle size separation at around 75 µm at mass throughput capacities of 3 tph/m^2 or 40 tph for the total unit, which incorporates five screen decks and 0.10-mm slot openings (Hollis 2007). Another screening system capable of high-capacity ultrafine separations is the Pansep screen, which was recently evaluated in a detailed test program by Mohanty et al. (2002). Although excellent screening efficiencies were achieved at a separation size of 45 µm, the undersize bypass increased to around 50% when the feed mass flow rate was increased above 2.6 tph/m^2 of screen area.

Classifying cyclones are the most commonly used technology for achieving size separations below 1 mm because of their high throughput capacity and the volume of slurry that typically exists in fine coal circuits. Recently, advancements in cyclone design have enhanced performance and resulted in improvements in operational characteristics such as lower operating pressures (Heiskanen 1996; Obeng and Morrell 2003; Rong and Napier-Munn 2003; Mohanty et al. 2002). However, inherent problems that have not been remedied include the amount of hydraulically entrained undersize material reporting to the coarse underflow stream and a density effect that results from the density difference between coal and mineral matter.

Based on the work of Firth and O'Brien (2003), an in-plant study was conducted to investigate the use of 15-cm-diameter gMax cyclones in a rougher-cleaner classifying circuit. The objective was to produce clean coal by removing the slime material in the feed. Although ultrafine bypass was reduced to values as low as 6%, the density effect shown in Figure 4 resulted in an increase in the ash content of the coarser size fractions, which was counter to the study objective (Honaker et al. 2006a). Currently, Krebs Engineering in collaboration with researchers at Virginia Tech are redesigning the apex washing system that has been available for several decades with the goal of eliminating ultrafine bypass while minimizing the effect on separation size. If the classification process is limited to one unit instead of a rougher-cleaner approach, the density effect should be minimized.

Froth Flotation

The general agreement about the importance of bubble size in flotation is that a bubble size range of around 0.8 mm provides optimum performance for most flotation systems. Low collision efficiencies between particles and bubbles are realized for larger bubbles, and smaller bubbles result in less adhesion between particles and bubbles. A range of bubble sizes provides maximum flotation recovery.

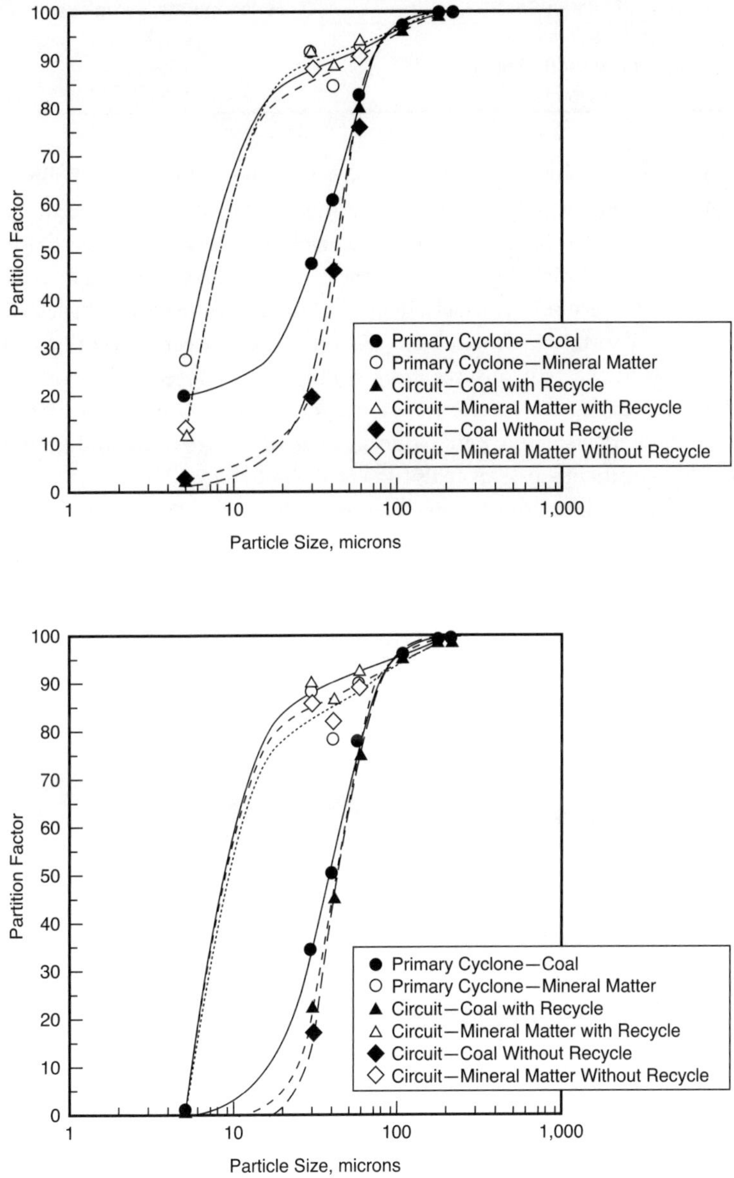

FIGURE 4 Coal and mineral matter size separation achieved by the classification circuit according to actual (top) and corrected (bottom) performance curves; primary apex = 0.50 in.

TABLE 2 Flotation rate values achieved by different feed conditioning methods

Flotation Test	Flotation Rate, per minute
Diesel fuel no. 2	0.657
Air eductor (test no. 1)	1.226
Air eductor (test no. 2)	1.259

However, recent studies have shown that the induction of bubbles of around 0.03 mm, which are called picobubbles, can enhance coal recovery by as much as 15% (Atalla et al. 2000). The picobubbles can be produced using a cavitation tube or ultrasonic device. If the feed coal slurry is treated in a cavitation tube, the picobubbles nucleate on the surface of the coal particle, which enhances the surface hydrophobicity. As a result, the larger conventional bubbles attach to the coal surfaces more readily (i.e., reduced induction time), thereby increasing the rate of flotation and overall recovery. Other benefits identified in a recent study are reduced collector dosage needs and improved recovery of difficult-to-float coals (Tao et al. 2006).

In a recent flotation project, flotation feed slurry from a central Appalachia coal preparation plant treating Coalberg seam coal was treated through an air eductor supplied by Eriez Manufacturing, and flotation rate tests were performed on the coal slurry after treatment. As shown in Table 2, the flotation rate of the −150 μm Coalberg coal was nearly doubled by treatment through the air eductor. In-plant trials on a conventional flotation bank in the same plant revealed a potential mass yield improvement of 10 absolute percentage points.

Dewatering

The recovery of fine coal is often limited by the inability to achieve acceptable moisture values from currently available dewatering technologies. Minimizing fine coal moisture is important for meeting the plant optimization criteria based on constant incremental inerts (moisture plus ash content), especially for plants providing coal to the steam market. As a result, a significant amount of research is being conducted in an effort to develop technologies or dewatering aids that provide moisture values which do not limit the recovery of fine coal. Some of the unique developments were reviewed by Yoon et al. (2006).

A recent development of a novel dewatering technology using hyperbaric centrifugation has indicated the potential of achieving moisture contents of around 10% for fine clean coal having a particle size below 1 mm (Asmatulu et al. 2005). The unit uses the pressure provided by a mechanically applied centrifugal field to drive water through a filter membrane that covers the circumference of a cylindrical container. The container includes sealed bearings that allow pressurized air to be used in dewatering. Centrifugal fields of up to 2,500 g provided product moisture contents of nearly 20% for Pittsburgh No. 8

coal with a top particle size of 1.18 mm, which was reduced to 8.1% by the use of pressurized air at 300 kPa.

Plate-and-frame filtration technology is known to provide the ability to dewater and recover more than 99% of the particles feeding the unit while providing low product moisture contents. Common problems with the units include high capital and operating costs. A new unit that addresses these problems is the Technicas Hidraulicas filter press. In-plant test programs were performed on Illinois No. 5 and No. 6 flotation concentrates with a nominal top size of 150 µm (Patwardhan et al. 2006). For the Illinois No. 5 flotation product, which contains 14% inherent moisture, a product moisture content of around 26%–28% was consistently achieved. Values approaching 21% moisture were achieved for the Illinois No. 6 material, which had a mean size of 75 µm. The capacity of a 36-plate unit was estimated to be 23 tph.

The steel belt filter is a novel dewatering technology that combines vacuum and pressure filtration principles to dewater fine coal. The fine coal slurry is first added onto a belt that has vacuum boxes installed on its bottom. The filter cake that is formed from vacuum filtration is pressed between the primary belt and a second belt that runs on top of the primary belt. Because steel belts are employed, hot air can be used to further reduce product moisture. An extensive test program performed on the unit achieved a product moisture content of 18.5% from a spiral and flotation product blend having a mean particle size of 400 µm (Mohanty et al. 2004, 2005). The maximum capacity on a prototype unit was 1.5 tph per meter of belt width.

Thickening and Clarification

The use of slurry impoundments for the storage of fine coal refuse slurry has attracted significant concern that has resulted in substantial difficulty in receiving permits for new or modified impoundments during the past decade. Alternatives that are being permitted with less difficulty include densification of the thickener underflow using belt presses and codisposal of the filter cake material with the coarse refuse and use of slurry cells. Belt presses have high maintenance requirements, operating costs, and sensitivity to operating conditions. Slurry cell systems have a low capacity because of the need for the water to drain or evaporate and thus are typically limited to plant capacities less than 600 tph.

Deep cone thickening is an alternative to belt filters that may be used to enhance slurry cell production and reduce the volume of waste being disposed. In a deep cone thickener, flocculated solids settle into a decreasing cross-sectional area, which increases compression on the thickened bed of solids. As a result, water is squeezed out of the agglomerates and rises upward. Multiple high-torque rakes placed along the height of the thickener help release water layers that build up and keep the thick, highly viscous solids moving downward.

The underflow rate is controlled to maintain a desired thickened bed level. A pilot-scale Eimco deep cone thickener was tested at a coal preparation plant operating in the central Appalachia coalfields where codisposal of coarse and fine waste was being practiced. The deep cone thickener was fed to the underflow stream from an adjacent conventional thickener. The test program consistently produced an underflow stream with solid concentrations between 50% and 55% by weight and an overflow stream that was nearly free of suspended solids (Parekh et al. 2006). The project resulted in a commercial installation at an Arch Coal preparation plant and a fine coal recovery operation in the same coalfields (Cook et al. 2007). The cited advantages over belt presses at the Arch Coal site were 80% lower chemical costs with similar product solid concentrations and fewer maintenance needs.

DRY COAL CLEANING RESEARCH

The use of air rather than water as a separation medium has been an area of focused research in the past decade. The main reason for the attention in this area is an increasing need to clean low-rank coals and to process coal from arid regions. In the United States, economical extraction of high-quality coal from central Appalachia coalfields results in a significant amount of out-of-seam contamination that could be removed by a dry cleaning technology before loading and hauling to an off-site wet cleaning facility.

Recent efforts by Allminerals Ltd. to modify the stump jig technology have resulted in the Allair jig (Kelley and Snoby 2002). The jig uses hindered settling and consolidated trickling action provided through the application of pulsed air, as shown in Figure 5a. The unit has been successfully used in several applications in and outside the United States for coal cleaning (Weinstein and Snoby 2007). The air jig is typically designed to treat coarse (50 × 13 mm) and fine (–13 mm) coal in separate units. For the same size unit, the coarse coal feed capacity is between 100 and 120 tph, whereas the fine coal capacity is 50 to 80 tph. The separation density typically is around 2.0 RD, with Ep values in the range of 0.2–0.3.

The FGX separator is dry, density-based separation technology using autogenous medium fluidized bed and table concentration principles, as shown in Figure 5b (Lu et al. 2003; Li and Yang 2006). This technology was the focus of a U.S. Department of Energy research project, which involved the testing of a 5-tph pilot-scale unit of the FGX separator at several coal operations throughout the United States (Honaker et al. 2007). Typical separation performances are shown in Figure 6. The subbituminous coal was a waste material derived from contamination with out-of-seam rock at a Powder River Basin operation. From a feed containing about 23.90% ash, the FGX separator generated products containing as little as 6.19% ash while recovering 82%

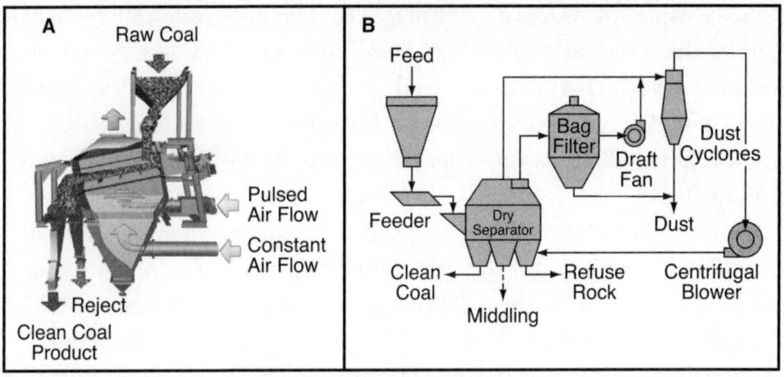

FIGURE 5 Schematics of (a) the Allair jig and (b) the FGX separator

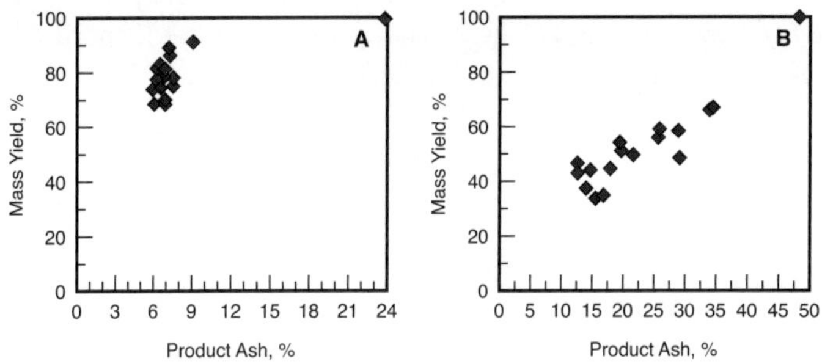

FIGURE 6 Separation performances achieved by the FGX separator on (a) subbituminous and (b) bituminous coal

of the feed mass. For the bituminous coal, the objective was to remove as much rock as possible while maintaining the recovery of 1.6 RD float material near 100%. The optimum performance was the rejection of 33.5% of the feed, which contained 88.3% ash. The result was a reduction in the coal ash content from 51.69% to 34.05%. Relative densities of separation were in the range of 1.8 to 2.2, with Ep values around 0.25. Although the technology has several successful commercial installations in China, the first U.S. installation is expected in 2008 and will process 120 tph of 50 × 13 mm lignite coal. The choice of a lignite installation is based on pilot-scale results indicating the ability to achieve 35% total sulfur reductions and 55% mercury reductions.

A significant development by Chinese researchers has been the air dense-medium fluidized bed (ADMFB), in which magnetite having a particle size as coarse as 500 μm is fluidized using air. The large magnetite particles are easier to recover using low-intensity dry magnets. The appearance of the separator resembles the standard shallow bath vessel that is popular in U.S. wet processing plants. The ADMFB cleans coal in the size range of 50 × 6 mm and achieves low-density separations with Ep values in the range of 0.05–0.07. Commercial installations with capacities up to 50 tph are in operation (Chen and Yang 2003).

Current research is focusing on the development of modifications needed to extend the application of the technology to particle sizes below 6 mm. One of the problems noted in the ADMFB process is a uniform distribution of air. Large voids of air periodically move through the fluidized bed, thereby inducing a significant amount of mixing. The mixing appears to have minimal effect on the performances achieved on the 50 × 6 mm fraction. However, the large air bubbles reduce the ability to achieve effective separation on –6 mm coal. To remedy this problem, Chinese researchers have applied a vibration frequency through the bed to minimize the size of the air voids moving through the bed (Chen and Yang 2003). A laboratory vibration-enhanced ADMFB unit has provided a product ash reduction from 16.57% to 8.35% while recovering 80.2% of the feed made up of 6 × 0.5 mm coal. The Ep value was 0.065.

In laboratory tests, Choung et al. (2006) found that similar separation efficiencies can be achieved on 6 × 1 mm coal without the need for mechanical vibration by simply decreasing the particle size of the magnetite. By using 75 × 45 μm magnetite, lower fluidization air rates can be used, which provides a more stable and uniform bed with less mixing. At a bed density of 1.6 RD, the ash content of 3 × 1 mm coal was reduced from 21% to 7%. The Ep values achieved by optimizing the magnetite size were below 0.10, with values for the 6 × 3 mm fraction as low as 0.03. Extending the application to cleaning 1 × 0.15 mm coal required reducing the top size of the magnetite to below 53 μm. However, process efficiencies significantly declined, as indicated by an Ep value of 0.10, because of an apparent increase in the viscosity of the fluidized bed. As a result, it was concluded that the minimum particle size for which the ADMFB can be applied to coal cleaning is 1 mm.

The most common dry cleaning technology for –1 mm coal is the triboelectric separator. Coal is tribocharged by contact with an electrically grounded material such as copper. Coal and mineral particles charge with positive and negative polarities, respectively. Afterwards, the charged particles are pneumatically transported between two oppositely charged plates or series of rolls to which oppositely charged particles are attracted. Stencel et al. (2002) found that coal and mineral matter particles are naturally charged in both the pressurized and suction configurations of utility pulverizers, and therefore potential exists for in-line removal of mineral matter before the combustion unit. Yoon et al. (1995) developed a continuous 250-kg/h triboelectric unit as part of a U.S. Department of Energy project. Test results indicated the ability to achieve 90% pyrite rejection and 70% ash rejection when treating fine coal. Recently, Tao and Jiang (2005) developed a novel tribocharge separator that uses a turbocharge mechanism rotating at speeds up to 10,000 rpm. The high-energy impact of the rotor with the particles results in significantly higher surface charge densities. A typical separation performance on −1 mm coal achieved an ash reduction from 19% to 9% while recovering 85% of the combustible material. According to Chen and Yang (2003), significant research effort into triboelectric separations is also being performed in China, where coal below 45 μm has been treated and product ash contents below 2% have been achieved.

SUMMARY AND CONCLUSIONS

A review of recent and current coal preparation research reveals that the majority of the effort is focused on dry coal cleaning and various aspects of water-based processing of fine coal. The interest in developing innovative, more efficient dry cleaning technologies for coarse and fine coal will continue to increase as a result of the growing need to process low-rank coals such as those in the Powder River Basin and the cleaning of coals in arid locations where water supply is limited. As new-generation combustion technologies are installed that allow fuels with higher ash content, dry coal cleaning may become more attractive as the best choice for unloading material handling systems and the combustion system.

Water-based processes will be the best choice for a majority of coal cleaning needs well into the foreseeable future because of their ability to achieve a wide range of separation densities while maintaining a high level of process efficiency. Circuitry will vary between locations based on the market demands for the coals being treated. Despite more than 100 years of research, there remains a significant need for research and development in the area of fine coal processing, especially involving ultrafine particle classification, froth flotation, dewatering, and handling. Recent research has resulted in an array of fine coal cleaning technologies that provide a high level of cleaning efficiency. Dewatering the fine product has improved significantly with the potential to achieve moisture levels of around 10% using innovative technologies and dewatering aids. However, additional research is needed to develop units and methods that provide economical and commercially practical solutions.

Although not previously discussed in this chapter, the innovations and developments in computational analysis have resulted in significant additional research reinvestigating the current design of coal cleaning units used in today's coal preparation plants. Computation fluid dynamic (CFD) analysis studies of classifying cyclones have already increased our understanding of the process and resulted in new cyclone designs. Improvements in CFD analysis are still needed to account for all interactions that occur in a typical process unit. However, CFD modeling and similar techniques are and will be the focus of numerous coal preparation studies.

REFERENCES

Abbott, J. 1981. The optimization of process parameters to maximize the profitability from a three-component blend. In *Proceedings of the First Australian Coal Preparation Conference*. New South Wales, Australia: The Coal Preparation Society. pp. 87–105.

Asmatulu, R., Luttrell, G.H., and Yoon, R.-H. 2005. Dewatering of fine coal using hyperbaric centrifugation. *Coal Prep.* 25(3):117–128.

Atalla, M., Chao, C., and Nicol, S.K. 2000. The role of cavitation in coal flotation. In *Proceedings of the Eighth Australian Coal Preparation Conference*. New South Wales, Australia: Australian Coal Preparation Society. pp. 337–350.

Bethell, P.J. 2002. Fine coal cleaning at Massey Energy. In *Proceedings of the 19th Annual International Coal Preparation Exhibition and Conference*, Lexington, KY. Stamford, CT: Primedia Business Exhibitions. pp. 1–14.

Bethell, P.J., and DeHart, G. 2006. Design construction and commissioning of the new 2000 TPH Arch Coal preparation plant. In *Proceedings of the XV International Coal Preparation Congress*, Vol. 1. Beijing: China University of Mining and Technology Press. pp. 79–88.

Chen, Q., and Yang, Y. 2003. Development of dry beneficiation of coal in China. *Coal Prep.* 23(1-2):3–12.

Choung, J., Mak, C., and Xu, Z. 2006. Fine coal beneficiation using an air dense medium fluidized bed. *Coal Prep.* 26(1):1–15.

Cierpisz, S., and Gottfried, B.S. 1977. Theoretical aspects of coal washer performance. *Int. J. Miner. Process.* 4:261–278.

Cook, B., Johnson, J., and Slottee, S. 2007. Prep plants adopt paste thickening for refuse disposal. *Coal Age* 112(5):56–60.

De Korte, G.J. 2002. Dense-medium beneficiation of fine coal revisited. In *Proceedings of the XIV International Coal Preparation Congress*. Johannesburg: Camera Press. pp. 43–46.

———. 2003. Comments on the use of tracers to test dense medium plant efficiency. *Coal Prep.* 23(5-6):251–266.

De Korte, G.J., Hand, P., and Forbes, A. 2002. Supertracer—A work in progress. In *Proceedings of the XIV International Coal Preparation Congress*. Johannesburg: Camera Press. pp. 483–488.

Firth, B., and O'Brien, M. 2003. Hydrocyclones circuits. *Coal Prep.* 23(4):167–183.

Galvin, K.P., Zhou, J., Belcher, B.D., Pratten, S.J., Callen, A.M., Lambert, N., and Nguyentranlam, G. 2004. *Investigation of the Reflux Classifier for Density and Size Separations—from Pilot to Full Scale*. Final Report. ACARP Report No. C11008. Queensland, Australia: Australian Coal Association Research Program.

Heiskanen, K.G.H. 1996. Developments in wet classifiers. *Int. J. Miner. Process.* 44:29–42.

Holcomb, T., Pinson, K., and Bethell, P.J. 2006. Expansion of the Pardee preparation plant from 500 to 750 tph through the installation of heavy media vessel and deslime column flotation circuits. In *Proceedings of the 23rd International Coal Preparation Exhibition and Conference*, Lexington, KY. Stamford, CT: Primedia Business Exhibitions. pp. 65–75.

Hollis, R. 2007. Blue Diamond Coal Company's Leatherwood preparation plant undergoes major circuit modification. In *Proceedings of the 24th International Coal Preparation Conference*, Lexington, KY. Stamford, CT: Primedia Business Exhibitions. pp. 95–112.

Honaker, R.Q., Boaten, F., Luttrell, G.H., Moorhead, R., and Coker, T. 2006a. Ultrafine coal classification using 15 cm gMax cyclone circuits. In *Proceedings of the 23rd International Coal Preparation Conference*, Lexington, KY. Stamford, CT: Primedia Business Exhibitions. pp. 19–36.

Honaker, R.Q., Jain, M., Parrekh, B.K., and Saracoglu, M. 2006b. Optimized spiral separation performance for ultrafine coal cleaning. In *Proceedings of the XV International Coal Preparation Congress*, Vol. 2. Beijing: China University of Mining and Technology Press. pp. 502–508.

Honaker, R.Q., Luttrell, G.H., Bratton, R., Saracoglu, M., Thompson, E., and Richardson, V. 2007. Dry coal cleaning using the FGX separator. In *Proceedings of the 24th International Coal Preparation Conference*, Lexington, KY. Stamford, CT: Primedia Business Exhibitions. pp. 61–76.

Kelley, M., and Snoby, R. 2002. Performance and cost of air jigging in the 21st century. In *Proceedings of the 19th Annual International Coal Preparation Exhibition and Conference*, Lexington, KY. Stamford, CT: Primedia Business Exhibitions. pp. 175–186.

Kohmuench, J.N., Mankosa, M.J., Honaker, R.Q., and Bratton, R.C. 2006. Applications of the CrossFlow teeter-bed separator in the U.S. coal industry. *Miner. Metall. Process. J.* 23(4):187–195.

Li, G., and Yang, Y. 2006. Development and application of FGX series compound dry coal cleaning system, China Coal. Technology Monograph of the Tangshan Shenzhou Machinery Co., Ltd. pp. 17–28.

Lu, M., Yang, Y., and Li, G. 2003. The application of compound dry separation technology in China. In *Proceedings of the 20th Annual International Coal Preparation Exhibition and Conference*, Lexington, KY. Stamford, CT: Primedia Business Exhibitions. pp. 79–95.

Mohanty, M.K., Palit, A., and Dube, B. 2002. A comparative evaluation of new fine particle size separation technologies. *Miner. Eng.* 15:727–736.

Mohanty, M.K., Huang, Z., Wang, Z., and Hirschi, J. 2004. Optimization of the dewatering performance of a steel belt filter. *Coal Prep.* 24(1-2):53–68.

Mohanty, M.K., Huang, Z., Wang, Z., and Hirschi, J. 2005. Steel belt filter: A suitable technology for dewatering fine clean coal and tailings. *Coal Prep.* 25(1):1–14.

Obeng, D.P., and Morrell, S. 2003. The JK three-product cyclone-performance and potential applications. *Int. J. Miner. Process.* 69:129–142.

Parekh, B.K., Patil, D., Honaker, R.Q., and Baczek, F. 2006. Improving the densification of fine coal refuse slurries to eliminate slurry ponds. In *Proceedings of the XV International Coal Preparation Congress*, Vol. 2. Beijing: China University of Mining and Technology Press. pp. 549–554.

Patwardhan, A., Chugh, Y.P., Arnold, B.J., and Terblanche, A.N. 2006. Dewatering ultrafine coal in a T.H. filter press. *Coal Prep.* 26(1):33–54.

Rong, R., and Napier-Munn, T.J. 2003. Development of a more efficient classifying cyclone. *Coal Prep.* 23:149–165.

Stencel, J.M., Schaefer, J.L., Ban, H., Li, T., and Neathery, J.K. 2002. Mineral matter rejection in-line between pulverizers and burners at a utility. In *Impact of Mineral Impurities in Solid Fuel Combustion*. Edited by R.P. Gupta, T.F. Wall, and L. Baxter. New York: Kluwer Academic. pp. 645–651.

Tao, D., and Jiang, X. 2005. Electrostatic particle charger, electrostatic separation system and related methods. U.S. Patent 2521917.

Tao, D., Yu, S., and Parekh, B.K. 2006. Picobubble enhanced fine coal flotation. In *Proceedings of the XV International Coal Preparation Congress*, Vol. 1. Beijing: China University of Mining and Technology Press. pp. 385–392.

Weinstein, R., and Snoby, R. 2007. Advances in dry jigging improves coal quality. *Min. Eng.* 59(1):29–34.

Yoon, R.-H., Luttrell, G.H., and Adel, G.T. 1995. POC-scale testing of a dry triboelectrostatic separator for fine coal cleaning. In *Proceedings of the 11th Annual Coal Preparation, Utilization and Environmental Control Contractors Conference*, Pittsburgh, PA, July 12–14. Washington, DC: U.S. Department of Energy. pp. 65–72.

Yoon, R.-H., Eryadin, M.K., Zhang, J., Keles, S., and Luttrell, G.H. 2006. Development of advanced fine coal dewatering technologies. In *Proceedings of the XV International Coal Preparation Congress*, Vol. 2. Beijing: China University of Mining and Technology Press. pp. 575–583.

Index

NOTE: *f.* indicates figure; *t.* indicates table.

T

U

V

W